Instructor's Manual and Testbank

for

Clinical Calculations

With Applications to General and Specialty Areas

Third Edition

JOYCE L. KEE, RN, MS

Associate Professor Emerita
College of Nursing
University of Delaware
Newark, Delaware

SALLY M. MARSHALL, RN, MSN

Nursing Service
Department of Veterans Affairs
Regional Office and Medical Center
Wilmington, Delaware

W. B. SAUNDERS COMPANY

A Division of Harcourt Brace & Company

Philadelphia London Toronto Montreal Sydney Tokyo

W. B. SAUNDERS COMPANY
A Division of Harcourt Brace & Company
The Curtis Center
Independence Square West
Philadelphia, Pennsylvania 19106-3399

NOTICE

Nursing is an ever-changing field. Standard safety precautions must be followed, but as new research and clinical experience broaden our knowledge, changes in treatment and drug therapy become necessary or appropriate. The editors of this work have carefully checked the generic and trade drug names and verified drug dosages to ensure that the dosage information in this work is accurate and in accord with the standards accepted at the time of publication. Readers are advised, however, to check the product information currently provided by the manufacturer of each drug to be administered to be certain that changes have not been made in the recommended dose or in the contraindications for administration. This is of particular importance in regard to new or infrequently used drugs. It is the responsibility of the treating physician, relying on experience and knowledge of the patient, to determine dosages and the best treatment for the patient. The editors cannot be responsible for misuse or misapplication of the material in this work.

THE PUBLISHER

Instructor's Manual and Testbank for
Clinical Calculations: With Applications to
General and Specialty Areas, Third Edition

ISBN 0-7216-5451-7

Printed in the United States of America

Last digit is the print number: 9 8 7 6 5 4 3 2 1

Preface

The *Instructor's Manual and Testbank for Clinical Calculations,* 3rd edition, was developed to assist the instructor in how to make best use of the textbook. Part 1 covers teaching strategies and offers three different course outlines for teaching calculation of drug dosage: (1) as a separate course, (2) as part of a pharmacology course, or (3) integrated into the nursing curriculum and covered in the skill laboratory and clinical courses.

Part 2 of this Manual is a Basic Math Test for the instructor to administer to the students early in the course in order to evaluate basic math skills.

The Testbank Questions in Part 3 include 289 drug calculation problems covering general and specialty areas. Many of the drug problems include drug labels. Specific areas covered include orals (tablets and capsules), oral suspensions, injectables, pediatrics, drug calculations by body surface area (BSA), and intravenous fluids (hours to administer, direct IV injection or IV push, IV drop rate, antibiotics, critical care, and Ob/Gyn IV drugs).

The instructor can use these drug problems to test the students' competency in calculating patient drug dosages correctly. These problems are conveniently divided into 11 subareas and can be used in any order that relates to the curriculum.

Finally, Part 4 contains answers to all of the testbank questions in Part 3.

The drug labels in this manual appear courtesy of

Wyeth-Ayerst Laboratories, Philadelphia, PA
DuPont Merck Pharmaceutical Company
Parke-Davis Company
Warner Chilcott Laboratories
SmithKline Beecham Pharmaceuticals
Ortho-McNeil Pharmaceuticals
Hoechst Marion Roussel, Inc.
Merck and Company, Inc.
Bristol-Myers Squibb Company
Eli Lilly and Company
Reproduced with permission of Burroughs Wellcome Co.
Copyright, Lederle Laboratories Division of American Cyanamid Company, All Rights Reserved. Reprinted with Permission.

Contents

PART ONE

Teaching Strategies

Teaching calculation of drug dosages can be accomplished in a number of ways: (1) as a separate course, (2) as part of a pharmacology course, (3) or integrated into the nursing curriculum with the skill and practice laboratory and clinical courses. A sample drug calculation education curriculum for each of these methods follows. These suggested methods can be modified according to your program's curriculum and course time allotment.

I. CLINICAL CALCULATIONS AS A SEPARATE COURSE

Session 1

a. Faculty and students briefly discuss the course, assignments, tests, practice time, and students' responsibilites.

b. Assignment: Part I: Basic Math Review

Session 2

a. Administer the Basic Math Test (Part Two of this Manual). A passing score is 90%, or 45 correct answers. For a score of less than 45, the student reviews Part I of the textbook and retakes the Basic Math Test or a similar test.

b. Assignments
   - ❑ Chapter 1: Systems Used for Drug Administration
   - ❑ Chapter 2: Conversion with the Metric, Apothecary, and Household Systems
   - ❑ Chapter 3: Interpretation of Drug Labels, Orders and Charting, and Abbreviations

   (Note: Alternatively, the Basic Math test can be given in the third session, and Chapter 1, 2, and 3 can be assigned during the first session.)

Session 3

a. Faculty and students discuss the following (1½–2 hours):
   - ❑ Metric, Apothecary, and Household systems used for drug administration
   - ❑ Conversion in Metric and Apothecary system according to weight and conversion in Metric, Apothecary, and Household systems by volume
   - ❑ Interpreting drug labels, drug orders and charting, methods of drug distribution, 5 Rights in drug administration, and abbreviations

b. Assignments:
   - ❑ Chapter 4: Alternative Methods for Drug Administration
   - ❑ Chapter 5: Methods of Calculation

Session 4

a. Quiz related to Chapters 1, 2, and 3 (½ hour)

b. Faculty and students discuss the following (1½ hours):
   - ❑ Alternative methods for drug administration
   - ❑ Methods of calculation. Faculty and students can decide on one method for calculating drug dosages.
   - ❑ Use of practice problems

c. Assignments:
- ❑ Chapters 6: Oral and Enteral Preparations with Clinical Applications
- ❑ Chapter 7: Injectable Preparations with Clinical Applications

Session 5

a. Quiz related to Chapters 4 and 5 (1/2 hour)

b. Faculty and students discuss the following (1 1/2 hours):
- ❑ Oral and enteral preparations of drugs
- ❑ Injectable preparations of drugs

c. Assignment: Laboratory time for students to practice oral and injection administration

Session 6

a. Provide additional laboratory time for students to practice oral and injection admininstration.

b. Assignment: Chapter 8: Intravenous Administration

Session 7

a. Administer test covering Chapters 6 and 7 (1 hour).

b. Faculty and students discuss calculations for intravenous therapy (Chapter 8) (1 hour).

c. Assignments:
- ❑ Laboratory time for students to practice set-up of intravenous sets and regulation of IV flow rates
- ❑ Chapter 9: Pediatrics

Session 8

a. Administer quiz or test covering Chapter 8 (1/2–1 hour)

b. Faculty and students discuss calculating drug doses by body weight and body surface area for children (1–1 1/2 hours).

c. Assignments (Note: Assign the following chapters at this time or during the curriculum time devoted to these topics.):
- ❑ Chapter 10: Critical Care
- ❑ Chapter 11: Pediatric Critical Care
- ❑ Chapter 12: Labor and Delivery
- ❑ Chapter 13: Community

Session 9

a. Administer quiz covering Chapter 9 (1/2 hour).

b. Faculty and students discuss chapters previously assigned.

c. Provide additional practice time for oral, injection, and IV administrations as needed.

Final Exam

Final exam for course (2 hours) covers the following:
- ❑ Chapter 6: Oral and Enteral Preparations with Clinical Applications
- ❑ Chapter 7: Injectable Preparations with Clinical Applications
- ❑ Chapter 8: Intravenous Administration
- ❑ Chapter 9: Pediatrics

II. CLINICAL CALCULATIONS WITH PHARMACOLOGY COURSE AND SKILL LABORATORY

Prior to First Class

a. Assignment: Part I: Basic Math Review

b. Administer the Basic Math Test (Part Two of this Manual) at a specified time later; a passing score is 90%, or 45 correct answers.

Weeks 1–3 of Pharmacology Course

a. Assignments:
- ❑ Chapter 1: Systems Used for Drug Administration
- ❑ Chapter 2: Conversion with the Metric, Apothecary, and Household Systems
- ❑ Chapter 3: Interpretation of Drug Labels, Orders and Charting, and Abbreviations
- ❑ Chapter 4: Alternative Methods for Drug Administration
- ❑ Chapter 5: Methods of Calculation

b. Arrange a small group discussion or a class period for a question and answer hour on the assigned chapters.

Weeks 4–6 of Pharmacology Course

a. Assignments:
- ❑ Chapter 6: Oral and Enteral Preparations with Clinical Applications
- ❑ Chapter 7: Injectable Preparations with Clinical Applications

b. Arrange a question and answer time in groups or in class to cover Chapters 6 and 7 (1–2 hours).

c. Provide laboratory time for students to practice administering oral and injectable drugs (2 hours).

d. Give a quiz on topics from Chapters 6 and 7, or include these questions in with the pharmacology test.

Weeks 7 and 8 of Pharmacology Course

a. Assignments:
- ❑ Chapter 8: Intravenous Preparations with Clinical Applications
- ❑ Chapter 9: Pediatrics

b. Arrange a question and answer time in groups or in class to cover Chapters 8 and 9 (1–2 hours).

c. Provide laboratory time for students to practice setting up IV sets and monitoring IV flow rates (1–2 hours).

d. Give a quiz on topics from Chapters 8 and 9, or include these questions in with the pharmacology test.

Final Exam

a. Final exam for course can cover Chapters 3, 4, 6, 7, 8, and 9, or this material can be included in with the pharmacology test.

b. Calculation of drugs for critical care (adult and pediatric), labor and delivery, and community can be covered when these subjects are presented in the curriculum.

III. CLINICAL CALCULATIONS INTEGRATED INTO THE NURSING CURRICULUM (with the skill and practice laboratory and specialty areas)

Prior to Laboratory Time

a. Assignments:
- ❑ Part I: Basic Math Review
- ❑ Chapter 1: Systems Used for Drug Administration
- ❑ Chapter 2: Conversion with the Metric, Apothecary, and Household Systems
- ❑ Chapter 3: Interpretation of Drug Labels, Orders and Charting, and Abbreviations
- ❑ Chapter 4: Alternative Methods for Drug Administration
- ❑ Chapter 5: Methods of Calculation

b. Answer questions related to these chapters in small group discussions or during an introductory class session.

c. Administer a test covering the Basic Math Review and the first five chapters at a specified time later.

During Practice Laboratory

a. Assignments:
- ❑ Chapter 6: Oral and Enteral Preparations
- ❑ Chapter 7: Injectable Preparations
- ❑ Chapter 8: Intravenous Preparations

b. Provide laboratory time for students to practice administering oral and injectable drugs.

d. Give written and practice test(s) on topics from Chapters 6–8 at designated times.

e. Drug calculations for specialty areas such as pediatrics, critical care (adult and pediatrics), labor and delivery, and community should be included when these subjects are presented in the curriculum.

Basic Math Test

This math test is composed of five sections: Roman and Arabic numerals, fractions, decimals, ratio and proportion, and percentage. It contains a total of 50 questions. A passing score is 45 or more correct answers (90%). A nonpassing score is 6 or more incorrect answers.

I. ROMAN AND ARABIC NUMERALS

Convert these Roman numerals to Arabic numerals.

1. viii __________
2. xix __________
3. xxii __________
4. xxiv __________
5. XLVI __________

Convert these Arabic numerals to Roman numerals.

6. 4 __________
7. 14 __________
8. 27 __________
9. 39 __________
10. 44 __________

II. FRACTIONS

Which of the fractions in each pair has the largest value?

11. 1/4 or 1/6 __________
12. 1/200 or 1/150 __________

Reduce these improper fractions to whole or mixed numbers.

13. 45/5 = __________
14. 58/12 = __________

Change this mixed number to an improper fraction.

15. 9 3/4 = __________

Change these fractions to decimals.

16. 3/4 = __________
17. 5/8 = __________

Multiply these fractions.

18. 3/5 × 3 2/3 = __________

Divide these fractions.

19. 1/4 ÷ 1/2 = __________
20. 7 3/4 ÷ 2 = __________

III. DECIMALS

Round off these decimals to tenths.

21. 0.82 = __________
22. 2.45 = __________
23. 28.67 = __________

Change these decimals to fractions.

24. 0.21 = __________
25. 0.45 = __________
26. 0.006 = __________

Multiply these decimals.

27. 3.12 × 0.45 = __________
28. 21.9 × 1.7 = __________

Divide these decimals.

29. 5.3 ÷ 0.23 = __________
30. 75.5 ÷ 0.67 = __________

IV. RATIO AND PROPORTION

Change these ratios to fractions.

31. 2:3 = __________
32. 1:150 = __________
33. 45:85 = __________

Solve these ratios for x.

34. 2:10::x:50 x = __________
35. 0.45:100::x:1000 x = __________
36. 9:1000::x:4500 x = __________
37. 2.5:200::5:x x = __________

Change these ratios to fractions and solve for x.

38. 3:12::x:48 x = __________
39. 4.5:100::x:400 x = __________
40. 0.75:300::5:x x = __________

V. PERCENTAGE

Change these percentages to fractions.

41. 0.45% = __________
42. 7.35% = __________
43. 27.4% = __________

Change these percentages to decimals.

44. 25% = __________
45. 6% = __________
46. 3.5% = __________
47. 0.75% = __________

Change these percentages to ratios.

48. 9% = __________
49. 2.5% = __________
50. 65.52% = __________

ANSWERS TO BASIC MATH TEST

1. 8
2. 19
3. 22
4. 24
5. 46
6. iv
7. xiv
8. xxvii
9. xxxix
10. LVIV or lviv
11. 1/4
12. 1/150
13. 9
14. 4.83
15. 39/4, or $\frac{39}{4}$
16. 0.75
17. 0.625
18. $\frac{\cancel{3}^{1}}{5} \times \frac{11}{\cancel{3}_{1}} = \frac{11}{5} = 2\frac{1}{5}$
19. $\frac{1}{\cancel{4}_{2}} \times \frac{\cancel{2}^{1}}{1} = \frac{1}{2}$
20. $\frac{31}{4} \times \frac{1}{2} = \frac{31}{8} = 3\frac{7}{8}$
21. 0.8
22. 2.5
23. 28.7
24. 21/100
25. 45/100
26. 6/1000
27. 1.404, or 1.4
28. 37.23
29. 23.04
30. 112.68
31. 2/3
32. 1/150
33. 45/85, or $\frac{45}{85} = \frac{9}{17}$
34. $10x = 100$
    $x = 10$
35. $100x = 450$
    $x = 4.5$
36. $1000x = 40500$
    $x = 40.5$
37. $2.5x = 1000$
    $x = 400$
38. $\frac{3}{12} \times \frac{x}{48}$
    $12x = 144$
    $x = 12$
39. $\frac{4.5}{100} \times \frac{x}{400}$
    $100x = 1800$
    $x = 18$
40. $\frac{0.75}{300} \times \frac{5}{x}$
    $0.75x = 1500$
    $x = 2000$
41. $\frac{0.45}{100}$ or $\frac{45}{10{,}000}$
42. $\frac{7.35}{100}$ or $\frac{735}{10{,}000}$
43. $\frac{27.4}{100}$ or $\frac{274}{1000}$
44. 0.25
45. 0.06
46. 0.035
47. 0.0075
48. 9:100
49. 2.5:100
50. 65.52:100

PART THREE

Testbank Questions: Calculations

TO THE INSTRUCTOR

This testbank includes 300 drug calculation problems related to administration of drugs from oral, injectable, and intravenous preparations and other drug problems related to specialty areas such as pediatrics and critical care. Drug labels accompany many of the drug calculation problems and must be used to solve them. Instructors can use any or all of these problems for testing their students' ability to solve drug dosage questions. Answers to these problems are given in Part Four of this manual.

ORALS (TABLETS AND CAPSULES)

1. Order: HydroDiuril 50 mg, PO, q.d.
   Drug available: (See drug labels.)

USUAL ADULT DOSAGE: 2 to 4 tablets daily. See accompanying circular. Keep container tightly closed. Protect from light, moisture, freezing, -20°C (-4°F) and store at room temperature, 15-30°C (59-86°F). Dispense in a tight container. CAUTION: Federal (USA) law prohibits dispensing without prescription. This is a bulk package and not intended for dispensing. 100 | No. 3263 7833217

NDC 0006-0042-68
100 TABLETS
HYDRODIURIL® 25 mg
(HYDROCHLOROTHIAZIDE)
Dist. by:
MERCK & CO., INC.
West Point, PA 19486, USA

HydroDiuril® 100 mg
(Hydrochlorothiazide)
Dist. by:
MERCK & CO., INC.
West Point, PA 19486, USA
Keep container tightly closed. Protect from light, moisture, freezing, -20°C (-4°F) and store at room temperature, 15-30°C (59-86°F).
100 Tablets
Lot
Exp.
0006-0410-68

NDC 0006-0410-68
USUAL ADULT DOSAGE: ½ or 1 tablet daily. See accompanying circular.
CAUTION: Federal (USA) law prohibits dispensing without prescription.
This is a bulk package and not intended for dispensing. Dispense in a tight container.

   a. Which bottle of HydroDiuril would you use?________________

   b. How many tablet(s) would you give? (Show your work.)

   ____

   ____

   ____

2. Order: Benadryl 12.5 mg, PO, stat
   Drug available:

Usual Dosage—Adults, 1 capsule three or four times daily, or as directed by the physician. See package insert.
Dispense in a tight container as defined in the USP.
The pink capsule with white band is a trademark registered in the United States Patent Office.
Exp date and lot
0373G094

N 0071-0373-24
KAPSEALS®
Benadryl®
(Diphenhydramine HCl Capsules, USP)
50 mg
Caution—Federal law prohibits dispensing without prescription.
100 CAPSULES
PARKE-DAVIS
People Who Care

Keep this and all drugs out of the reach of children.
Store at controlled room temperature 15° to 30°C (59° to 86°F). Protect from moisture.
©1992, Warner-Lambert Co.
PARKE-DAVIS
Div of Warner-Lambert Co
Morris Plains, NJ 07950 USA
0071-0373-24

Dose—Adults, 1 or 2 capsules; Children, 1 capsule; three or four times daily, or as directed by the physician. See package insert.
Keep this and all drugs out of the reach of children.
Dispense in a tight container as defined in the USP.
Store at controlled room temperature 15°-30°C (59°-86°F).
Protect from moisture.
Capsule P-D 471 for prescription dispensing only.
Exp date and lot
0471G054

N 0071-0471-24
Benadryl®
(Diphenhydramine HCl Capsules, USP)
25 mg
Caution—Federal law prohibits dispensing without prescription.
100 CAPSULES
PARKE-DAVIS
People Who Care

©1992, Warner-Lambert Co.
PARKE-DAVIS
Div of Warner-Lambert Co
Morris Plains, NJ 079
0071-0471-24

   a. Which bottle of Benadryl would you use?________________

   b. How many capsule(s) would you give?________________

   c. Explain your answer: ________________

   ____

   ____

3. Order: Aspirin gr X, PO, b.i.d.
Drug available:

N 0047-0606-32

Aspirin Tablets, USP

Analgesic (Pain Reliever)/ Antipyretic (Fever Reducer)

Directions—Adults: Oral dosage is 1 tablet every three hours; or 1 to 2 tablets every four hours; or 2 to 3 tablets every six hours, while symptoms persist, not to exceed 12 tablets in any 24-hour period, or as directed by a doctor. Drink a full glass of water with each dose. **Children under 12 years of age:** Consult a doctor.

Indications—For the temporary relief of minor aches and pains and to reduce fever.

Quality Sealed for your protection*

*Do not use if the innerseal over the opening of the bottle printed "SEALED for YOUR PROTECTION" is broken or missing.

1000 Tablets
325 mg (5 grains) each

6505-00-153-8750

Active Ingredient—Each tablet contains Aspirin, USP ... 325 mg. Also contains corn starch and microcrystalline cellulose.
Warnings—Children and teenagers should not use this medicine for chicken pox or flu symptoms before a doctor is consulted about Reye syndrome, a rare but serious illness reported to be associated with aspirin.
Keep this and all drugs out of the reach of children. In case of accidental overdose, seek professional assistance or contact a poison control center immediately.
As with any drug, if you are pregnant or nursing a baby, seek the advice of a health professional before using this product. IT IS ESPECIALLY IMPORTANT NOT TO USE ASPIRIN DURING THE LAST 3 MONTHS OF PREGNANCY UNLESS SPECIFICALLY DIRECTED TO DO SO BY A DOCTOR BECAUSE IT MAY CAUSE PROBLEMS IN THE UNBORN CHILD OR COMPLICATIONS DURING DELIVERY.
Do not take this product for pain for more than 10 days or for fever for more than 3 days unless directed by a doctor. If pain or fever persists or gets worse, if new symptoms occur, or if redness or swelling is present, consult a doctor because these could be signs of a serious condition.
Do not take this product if you are allergic to aspirin or if you have asthma, or if you have stomach problems (such as heartburn, upset stomach, or stomach pain) that persist or recur, or if you have ulcers or bleeding problems, unless directed by a doctor. If ringing in the ears or a loss of hearing occurs, consult a doctor before taking any more of this product.
Drug Interaction Precaution: Do not take this product if you are taking a prescription drug for anticoagulation (thinning the blood), diabetes, gout, or arthritis unless directed by a doctor.
Store below 30°C (86°F). Protect from moisture.

Exp date and lot

N 3 0047-0606-32 1

WARNER CHILCOTT LABS
Div of Warner-Lambert Co
Morris Plains NJ 07950 USA
©1991
0606G022

a. Gr X is equal to how many milligrams? (Use table if needed.)________________

b. How many tablet(s) would you give?________________

4. Order: Amoxicillin clavulanate (Augmentin) 0.5 g, PO, q8h
Drug available:

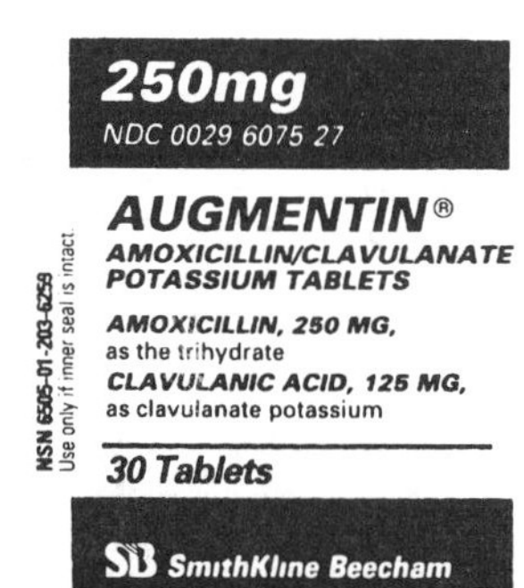

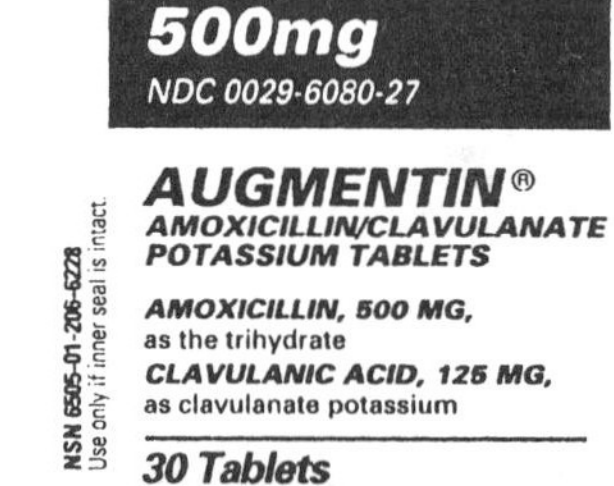

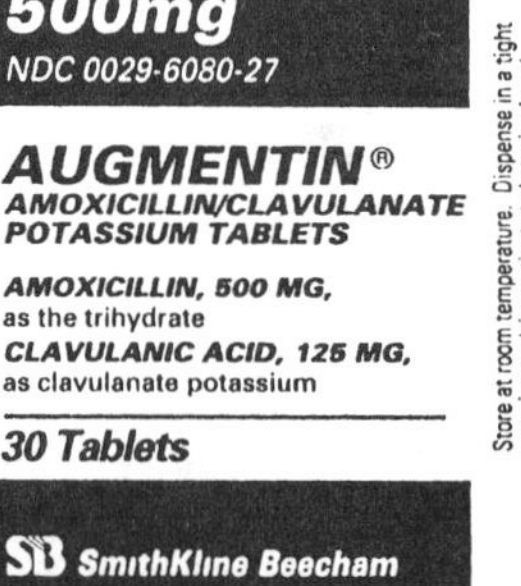

a. Which bottle would you use?________________

b. How many tablet(s) would you give?________________

11. Order: Cloxacillin 1 g/d, PO, in four divided doses, q6h
    Drug available:

a. How many milligrams should the patient receive q6h? ______________________

b. How many capsule(s) would you give per dose? ______________________

12. Order: Cardizem SR 120 mg, PO, b.i.d.
    Drug available:

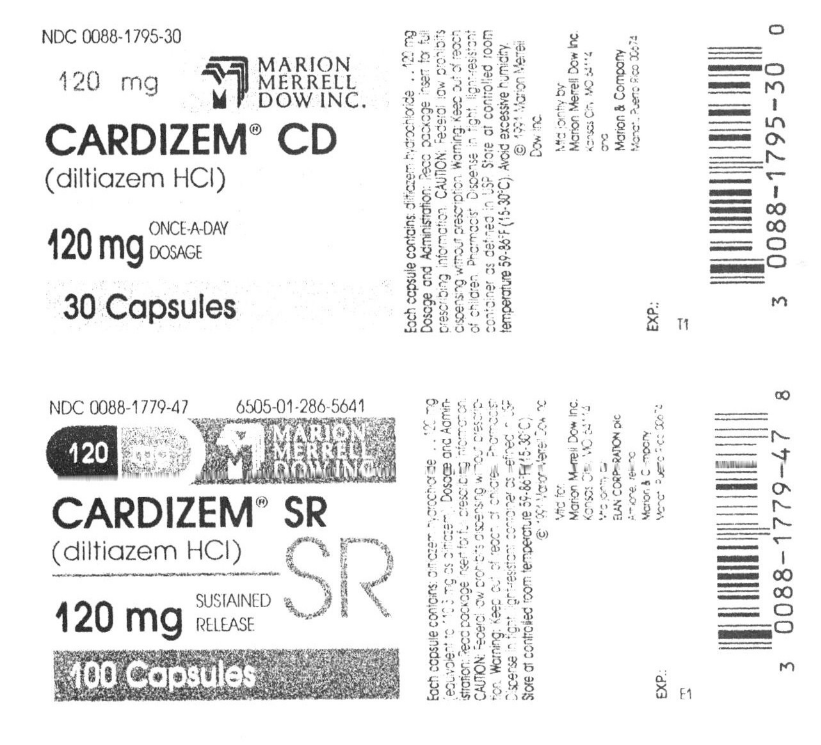

a. Which bottle of Cardizem would you use? ______________________

b. How many capsule(s) per day would the patient receive? ______________________

c. How many capsule(s) per dose would you give? ______________________

13. Order: Decadron 0.5 mg, PO, t.i.d.
    Drug available:

USUAL ADULT DOSAGE: See accompanying circular.
This is a bulk package and not intended for dispensing.
Dispense in a well-closed container.
CAUTION: Federal (USA) law prohibits dispensing without prescription.
100 | No. 7592 7834109

NDC 0006-0020-68
100 TABLETS
Decadron® 0.25 mg
(Dexamethasone)
Dist. by:
MERCK & CO., INC.
West Point, PA 19486, USA

Lot Exp.

a. How many milligrams would the patient receive per day?____________________

b. How many tablet(s) would you give per dose?____________________

____

____

14. Order: Coumadin 7.5 mg, PO, q.d.
    Drug available:

a. How many tablet(s) would you give?

____

____

____

*Present as crystalline sodium warfarin isopropanol clathrate. Store at controlled room temperature (59°-86°F, 15°-30°C).
USUAL ADULT DOSAGE: Read accompanying product information.
Dispense in a tight, light-resistant container as defined in the USP.
RESEAL CAP TIGHTLY.

Lot:
Exp:

NDC 0056-0176-90
NSN 6505-00-728-2617

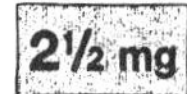

COUMADIN®
(Warfarin Sodium Tablets, USP)
Crystalline*

HIGHLY POTENT ANTICOAGULANT

WARNING: Serious bleeding results from overdosage. Do not use or dispense before reading directions and warnings in accompanying product information.

PROTECT FROM LIGHT. STORE IN CARTON UNTIL CONTENTS HAVE BEEN USED.

1000 TABLETS

CAUTION: Federal law prohibits dispensing without prescription.

DuPont Pharma
Wilmington, Delaware 19880

DU PONT PHARMA

9739/JC

15. Order: Procainamide (Procan) SR 0.5 g, PO, q6h
    Drug available:

Usual Dosage—See package insert for complete prescribing information.
Do not chew tablets.
Keep this and all drugs out of the reach of children.
Dispense in a tight container as defined in the USP.
Store below 30°C (86°F).
Protect from moisture.
Exp date and lot

6505-01-106-5973

0204G017

N 0071-0204-24
Procan® SR
(Procainamide Hydrochloride Extended-release Tablets, USP)
500 mg
Caution—Federal law prohibits dispensing without prescription.
100 TABLETS

PARKE-DAVIS
People Who Care

© 1992, Warner-Lambert Co.

Note: The drug in Procan SR tablets is 'held' in a wax core that has been designed to slowly release the drug into your system. When this process is completed, the empty wax core is eliminated from your body. Do not be concerned if you occasionally notice something that looks like a tablet in your stool.

PARKE-DAVIS
Div of Warner-Lambert Co
Morris Plains, NJ 07950 USA

N 0071-0204-24

a. How many grams (g) should the patient receive per day?____________________

b. How many tablet(s) of Procan will you give per dose?____________________

16. Order: Mandelamine 1 g, PO, q12h
    Drug available:

Note: Different tablet shape.
Each tablet contains:
Methenamine
Mandelate .... 0.5 gm
Usual Dosage—Adults:
Two tablets 4 times a day.
Children 6-12: One tablet
4 times a day.
Each brown tablet bears the
product code 166.
For full prescribing information
see package insert.
Keep this and all drugs out
of the reach of children.
Dispense in a tight container
as defined in the USP.
Reclose container tightly
with cap.
Store at controlled room
temperature 15°-30°C
(59°-86°F).
0166G160 LA45000 / H3

N 0071-0166-24
Mandelamine®
Hafgrams®
(Methenamine Mandelate
Tablets, USP)
0.5 gram
Film Coated
Caution—Federal law prohibits
dispensing without prescription.
100 TABLETS
PARKE-DAVIS
People Who Care

Manufactured by: ABLE LABORATORIES, INC.
South Plainfield, NJ 07080
For: PARKE-DAVIS
Div of Warner-Lambert Co ©1992
Morris Plains, NJ 07950 USA
Manufacturer's Code 53265

N 3 0071-0166-24 0
Lot No.:
Exp. Date:

    a. How many tablet(s) would you give?________________

17. Order: Methydopa (Aldomet) 0.75 g, PO, b.i.d.
    Drug available:

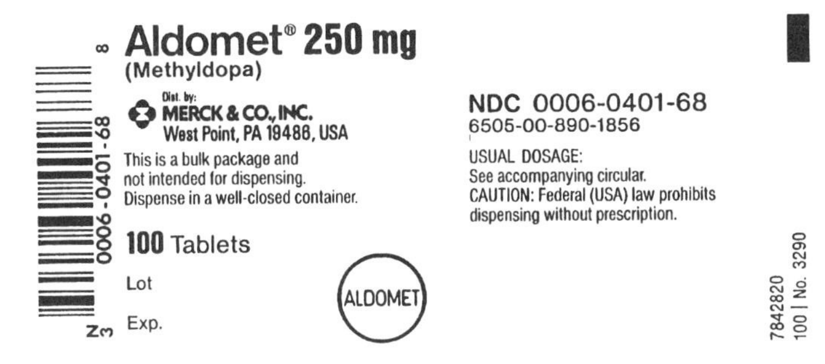

    a. How many grams of Aldomet would the patient receive per day?

    ____

    b. How many milligrams per day? ________________

    c. How many tablet(s) would you give per dose ?________________

18. Order: Digoxin (Lanoxin) 0.5 mg, PO, q.d.
    Drugs available:

    a. Which bottle of Digoxin (Lanoxin) would you use?________________

    b. How many tablet(s) would you give? ________________

19. Order: Kanamycin 0.5 g, PO, q12h
    Parameter: 15 mg/kg/d in two or three divided doses.
    Patient weighs 68 kg.
    Drug available:

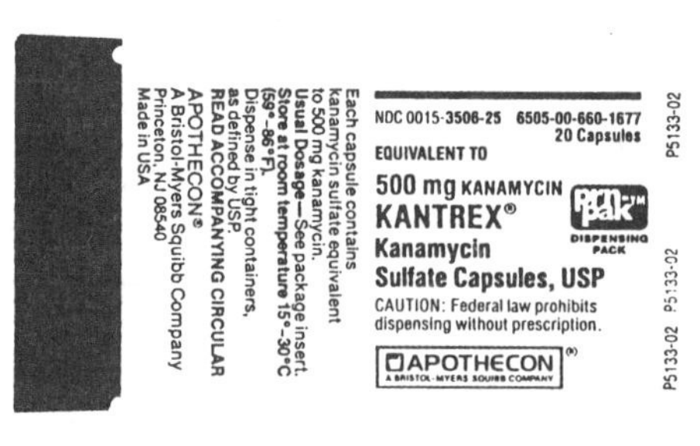

a. Is the patient's daily dose within safe dose parameters? Explain.____________________

____

b. How many tablet(s) should you give per dose? ____________________

20. Order: Cefuroxime (Ceftin) 0.25 g, PO, b.i.d.
    Drug available: Cefuroxime 125 mg tablets

a. How many milligrams equal 0.25 g?____________________

b. How many tablet(s) should you give per dose?____________________

21. Order: Etodolac 1.2 g/d, PO, in three divided doses
    Drug available: Etodolac 200 mg capsules

a. How many milligrams should the patient receive per dose?____________________

b. How many capsules would you give per dose?____________________

22. Order: Compazine 10 mg, PO, q.i.d.
    Drug available:

    (Maximum dose is 40 mg/d.)

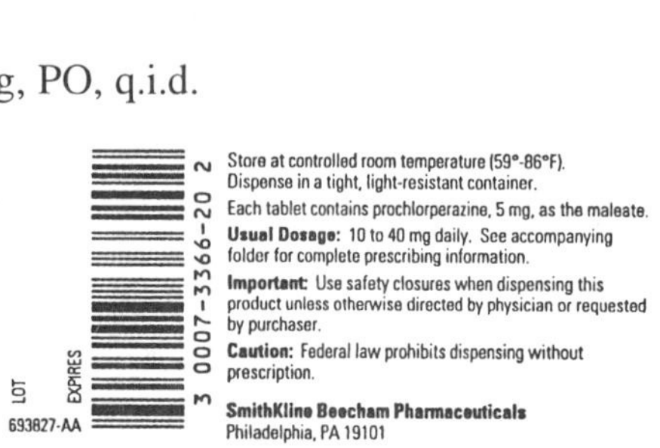

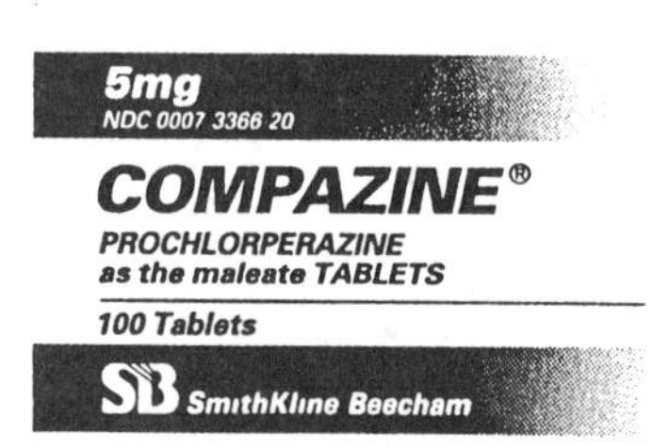

a. Is the dose per day within safe parameters? Explain. ____________________

____

b. How many tablet(s) would you give per dose? ____________________

23. Order: Glyburide 2.5 mg, PO, q.d.
    Drug available: Glyburide 1.25 mg and 5 mg

a. Which bottle of Glyburide would you use?____________________

b. How many tablet(s) of Glyburide would you give?____________________

24. Order: Prazosin 4 mg, PO, t.i.d.
    Drug available: Prazosin 1 mg, 2 mg, and 5 mg caplets

    a. Which bottle of Prazosin would you use?________________

    b. How many capsule(s) would you give per dose?________________

25. Drugs available:

Warning—To prevent loss of potency, keep these tablets in the original container. Close tightly immediately after each use.
N 0071-0569-24
Nitrostat®
(Nitroglycerin Tablets, USP)
0.3 mg (1/200 gr)
Caution—Federal law prohibits dispensing without prescription.
100 SUBLINGUAL TABLETS
PARKE-DAVIS
People Who Care
Usual Dosage—0.3 to 0.8 mg sublingually as needed.
See package insert for full prescribing information.
Keep this and all drugs out of the reach of children.
Dispense in original, unopened container.
Store at controlled room temperature 15°-30°C (59°-86°F).
Protect from moisture.
US Patent 3,789,119
PARKE-DAVIS
Div of Warner-Lambert Co ©1992
Morris Plains, NJ 07950 USA
0569G065

Warning—To prevent loss of potency, keep these tablets in the original container. Close tightly immediately after each use.
N 0071-0571-24
Nitrostat®
(Nitroglycerin Tablets, USP)
0.6 mg (1/100 gr)
Caution—Federal law prohibits dispensing without prescription.
100 SUBLINGUAL TABLETS
PARKE-DAVIS
People Who Care
Usual Dosage—0.15 to 0.6 mg sublingually as needed.
See package insert for full prescribing information.
Keep this and all drugs out of the reach of children.
Dispense in original, unopened container.
Store at controlled room temperature 15°-30°C (59°-86°F).
Protect from moisture.
US Patent 3,789,119
PARKE-DAVIS
Div of Warner-Lambert Co ©1992
Morris Plains, NJ 07950 USA
0571G044

Warning—To prevent loss of potency, keep these tablets in the original container. Close tightly immediately after each use.
N 0071-0570-24
Nitrostat®
(Nitroglycerin Tablets, USP)
0.4 mg (1/150 gr)
Caution—Federal law prohibits dispensing without prescription.
100 SUBLINGUAL TABLETS
PARKE-DAVIS
People Who Care
Usual Dosage—0.15 to 0.6 mg sublingually as needed.
See package insert for full prescribing information.
Keep this and all drugs out of the reach of children.
Dispense in original, unopened container.
Store at controlled room temperature 15°-30°C (59°-86°F).
Protect from moisture.
US Patent 3,789,119
PARKE-DAVIS
Div of Warner-Lambert Co ©1992
Morris Plains, NJ 07950 USA
0570G035

    a. Which of these nitroglycerin preparations is most potent?________________

    b. How is Nitrostat SL administered?________________

    Why?________________

    ____

26. Order: Lorazepam (Ativan) 1 mg, PO, t.i.d.
    Drug available:

NDC 0008-0081-02
Ativan®
(lorazepam)
0.5 mg
100 tablets
SEALED FOR YOUR PROTECTION
Keep tightly closed—Store at controlled room temperature. Dispense in tight container—The appearance of these tablets is a trademark of Wyeth Laboratories.
Made and printed in USA U0081-02-6
WYETH LABORATORIES INC.
Philadelphia PA 19101
Usual Dosage: One to six mg per day given in divided doses. See accompanying information. Caution: Federal law prohibits dispensing without prescription.
LOT
EXP

    a. How many milligrams of Ativan would the patient receive per day?________________

    b. How many tablet(s) would you give per dose?________________

27. Order: Ethambutol HCl 800 mg, PO, q.d.
    Parameter: 15 mg/kg/d
    Patient weighs 110 pounds
    Drug available: Ethambutol 100 mg and 400 mg

    a. How many kilograms does the patient weigh?________________

    b. What is the maximum dose the patient would receive per day? ________________

    c. Which bottle of Ethambutol would you use?________________

    d. How many tablet(s) would you give?________________

28. Order: Aminocaproic acid (Amicar) 1.5 g, PO, stat, and may repeat in 1 hour
Drug available:

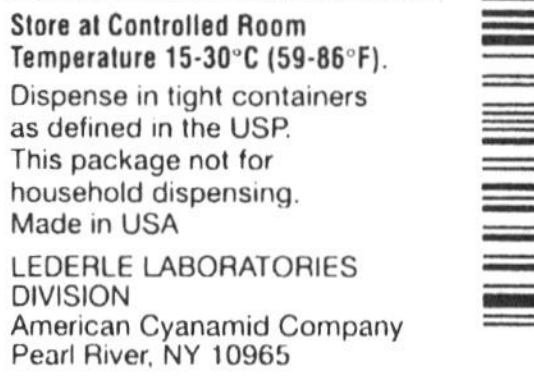

a. How many milligrams are in 1.5 g?____________________

b. How many tablet(s) would you give?____________________

29. Order: Ciprofloxacin (Cipro) 0.25 g, PO, q12h
Drug available: Cipro 250 mg and 500 mg tablets

a. Which bottle of ciprofloxacin would you choose?____________________

b. How many tablet(s) would you give?____________________

30. Order: Codeine gr 1, PO, stat
Drug available:

(Convert grains to milligrams; use table if needed.)

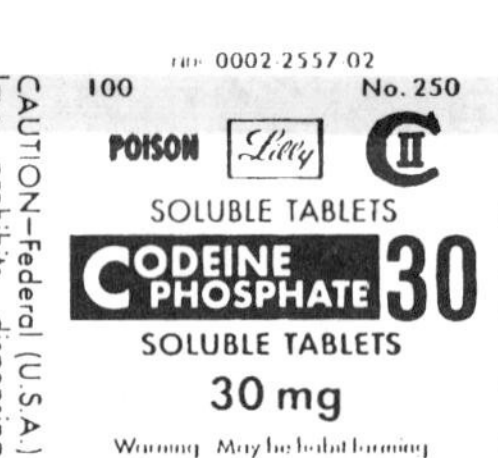

a. How many milligrams are in 1 gr?____________________

b. How many tablet(s) of codeine would you give? ____________________

31. Order: Propranolol (Inderal) 30 mg, PO, t.i.d.
Drug available: Propranolol 10 mg, 20 mg, 40 mg, 60 mg, and 80 mg

a. Which bottle(s) of propranolol would you use? ____________________

b. How many tablet(s) would you give per dose?____________________

32. Order: Tagamet 0.4 g, PO, b.i.d., and 0.8 g, at hour of sleep

Drugs available:

NSN 6505-01-103-6335
Store at controlled room temperature (59° to 86°F). Dispense in a tight, light-resistant container.
Each tablet contains cimetidine, 200 mg
Dosage: See accompanying prescribing information.
Important: Use safety closures when dispensing this product unless otherwise directed by physician or requested by purchaser
Caution: Federal law prohibits dispensing without prescription.
SmithKline Beecham Pharmaceuticals
Philadelphia, PA 19101
3 0108-5012-20 0
EXP

200mg
NDC 0108-5012-20
TAGAMET®
CIMETIDINE TABLETS
100 Tablets
SB SmithKline Beecham

NSN 6505-01-176-0712
Store at controlled room temperature (59° to 86°F).
Dispense in a tight, light-resistant container.
Each Tiltab® tablet contains cimetidine, 400 mg.
Dosage: See accompanying prescribing information.
Important: Use safety closures when dispensing this product unless otherwise directed by physician or requested by purchaser.
Caution: Federal law prohibits dispensing without prescription.
SmithKline Beecham Pharmaceuticals
Philadelphia, PA 19101
3 0108-5026-18 4
EXP

400mg
NDC 0108-5026-18
TAGAMET®
CIMETIDINE TABLETS
60 TILTAB® Tablets
SB SmithKline Beecham

a. Which bottle of Tagamet would you use?__________

b. How many tablet(s) would you give per dose during the day?__________

How many tablet(s) at hour of sleep? __________

33. Order: Cinoxacin 1 g/d, PO, in two divided doses
Drugs available: Cinoxacin 250 mg and 500 mg capsules

a. What are the specific times for the patient to receive cinoxacin?__________

b. How many grams or milligrams per dose?__________

c. Which bottle of cinoxacin would you use? __________

d. How many capsule(s) would you give? __________

34. Order: Captopril (Capoten) 50 mg, PO, b.i.d.
Drugs available: Captopril 12.5 mg, 25 mg, and 37.5 mg tablets

a. Which bottle of captopril would you use? __________

b. How many tablet(s) would the patient receive per dose?__________

35. Order: Betamethasone (Celestone) 2.4 mg, PO, q.d.
Drug available: Betamethasone 0.6 mg tablets

a. How many tablet(s) would you give?__________

36. Order: Dexamethasone (Decadron) 3 mg, PO, b.i.d.
Drugs available:

USUAL ADULT DOSAGE: See accompanying circular.
Dispense in a well-closed container.
CAUTION: Federal (USA) law prohibits dispensing without prescription.
50 | No. 7638 7834211
NDC 0006-0095-50
50 TABLETS
Decadron® 1.5 mg
(Dexamethasone)
Dist. by: MERCK & CO., INC.
West Point, PA 19486, USA
DECADRON

USUAL ADULT DOSAGE: See accompanying circular.
Dispense in a well-closed container.
CAUTION: Federal (USA) law prohibits dispensing without prescription.
50 | No. 7648 7834305
NDC 0006-0147-50
50 TABLETS
Decadron® 6 mg
(Dexamethasone)
Dist. by: MERCK & CO., INC.
West Point, PA 19486, USA
DECADRON

a. Which bottle of Decadron would you use?__________

b. How many tablet(s) would you give per dose?__________

37. Order: Dilantin 0.1 g, PO, q.d.
    Drugs available: Phenytoin (Dilantin) 30 mg and 100 mg capsules

    a. Which bottle of Dilantin would you use?__________
    b. How many capule(s) of Dilantin would you give?__________

38. Order: Phenytoin (Dilantin) 1 g, PO, loading dose (LD) in three divided doses in 24 hr
    Parameter: 15–18 mg/kg/LD
    Patient weighs 60 kg
    Drugs available: Dilantin 100 mg capsule

    a. In 24 hr, three divided doses would be every __________
    b. How many milligrams would the patient receive per dose in 24 hr? __________
    c. Is the dose within safe parameters?__________
       Explain: __________
    d. How many capsule(s) would you give per dose? __________

39. Order: Meprobamate (Equanil) 1.2 g/d, in three divided doses
    Drugs available: Equanil 200 mg and 400 mg tablets

    a. How many milligrams would the patient receive per dose?__________
    b. Which bottle of meprobamate would you use ?__________
    c. How many tablet(s) would you give per dose? __________

40. Order: Cyclophosphamide (Cytoxan) 200 mg, PO, q.d.
    Parameter: 1–5 mg/kg/d
    Patient weighs 100 pounds
    Drug available:

    100 TABLETS NDC 0015-0503-01
    CYTOXAN®
    (cyclophosphamide tablets, USP)
    50 mg
    CAUTION: FEDERAL LAW PROHIBITS DISPENSING WITHOUT PRESCRIPTION
    CYTOTOXIC AGENT
    Storage at or below 77° F (25° C) is recommended
    NSN 6505-00-733-5246
    Mead Johnson ONCOLOGY PRODUCTS

    a. How many kilograms does the patient weigh?__________
    b. Is the dose of Cytoxan within the parameter range? __________
       Explain:__________
    c. How many tablet(s) would you give?__________

ORAL SUSPENSIONS

41. Order: Dilantin 100 mg, PO, q.d.
    Drug available: Dilantin 250 mg per 5 mL

    a. How many milliliters would you give?____________________

42. Order: Amoxicillin 200 mg, PO, q8h
    Drug available:

    a. How many milliliters of Amoxicillin would you give per dose?

    ____

Amoxicillin 80 mL, 250 mg per 5 mL
Lederle
Keep tightly closed.
Shake well before using.
Bottle is not full to allow for shaking.
Store under refrigeration.
Discard any unused portion after 14 days.

Control
Exp

NDC 0005-3147-43
Amoxicillin
for oral suspension
250 mg
amoxicillin as the trihydrate per 5 mL
1 bottle powder (4 gram amoxicillin) to make 80 mL liquid.
USUAL DOSAGE: Adults - 250 mg to 500 mg every 8 hours. Children - 20 to 40 mg/kg/day in divided doses every 8 hours, depending on age, weight and severity of infection.
See accompanying information.
CAUTION: Federal law prohibits dispensing without prescription.
80 mL when reconstituted

Keep at Room Temperature [Approx. 25°C (77°F)] before reconstitution.
DIRECTIONS FOR RECONSTITUTION:
Reconstitute with 58 mL of water. Add the water in two separate portions, shaking well after each addition. The bottle will then contain 80 mL of suspension. Each 5 mL will contain amoxicillin trihydrate equivalent to 250 mg amoxicillin.

PULL

BOTTLE SEALED WITH PRINTED PLASTIC BAND
Tamper Resistant
IF BAND IS BROKEN DO NOT ACCEPT

N 3 0005-3147-43 2

14371-91 WY3 U3147-43-3
Manufactured by WYETH LABORATORIES INC., Philadelphia, PA 19101 for LEDERLE LABORATORIES DIVISION American Cyanamid Company, Pearl River, NY 10965

43. Order: Artane 1 mg, PO initially
    Drug available:

    a. How many milliliters would you give initially?____________________

    ____

Lederle NDC 0005-4440-65
Artane®
Trihexyphenidyl Hydrochloride
Elixir
This package not for household dispensing.
EACH TEASPOONFUL (5 mL) CONTAINS:
Trihexyphenidyl HCl 2 mg
Alcohol 5%
Preservatives:
Methylparaben 0.08%
Propylparaben 0.02%
AVERAGE DOSAGE:
3 to 5 teaspoonfuls (15-25 mL) daily for maintenance.
See Accompanying Literature.
CAUTION: Federal law prohibits dispensing without prescription.
Store at Controlled Room Temperature 15-30°C (59-86°F).
DO NOT FREEZE
Dispense in tight containers as defined in the USP.
Control No. Exp. Date

44. Order: Artane 5 mg, PO, b.i.d.
    Drug available: (same as in question #43)

    a. How many milliliters of Artane would you give per dose? ____________________

    ____

45. Order: Minocycline (Minocin) 100 mg, PO, q12h
    Drug available:

    a. How many milliliters would you administer per dose? ____________________

    ____

Lederle NDC 0005-5313-56
Minocin®
Minocycline Hydrochloride
Oral Suspension
50 mg per 5 mL
(Custard Flavored)
CAUTION: Federal law prohibits dispensing without prescription.
Usual Daily Dose for Children above 8 years of age: 4 mg/kg initially followed by 2 mg/kg every 12 hours.
2 FL. OZ. (60 mL) 21411-92 D6

SHAKE WELL
WARNING: NOT FOR INJECTION
Each teaspoonful (5 mL) contains Minocycline hydrochloride equivalent to 50 mg minocycline.
Propylparaben 0.10%
Butylparaben 0.06%
Alcohol USP 5.0% v/v
See accompanying circular.
Store at Controlled Room Temperature 15-30°C (59-86°F). DO NOT FREEZE.
LEDERLE LABORATORIES DIVISION
American Cyanamid Company
Pearl River, NY 10965

Control No.
Exp. Date

N 3 0005-5313-56 9

46. Ampicillin (Principen) 150 mg, PO, q6h
Drug available:

NDC 0003-0969-61
200 mL
NSN 6505-01-038-4540
EQUIVALENT TO
125 mg per 5 mL
when reconstituted according to directions.
PRINCIPEN®
Ampicillin for Oral Suspension, USP
CAUTION: Federal law prohibits dispensing without prescription.
APOTHECON® A BRISTOL-MYERS SQUIBB COMPANY

Control:
Exp. Date of powder: 12/1/98

Pharmacist: See base label for dispensing directions. Physician leaflet enclosed. Remove before dispensing.
READ ACCOMPANYING CIRCULAR
Usual Dosage: Adults—250 to 500 mg every 6 hours.
Children—50 to 100 mg/kg/day in divided doses.
Distributed by APOTHECON® A Bristol-Myers Squibb Company Princeton, NJ 08540 Made in USA 96961
79886OECL-2
++30003096961 1
OPEN ALONG PERFORATION
Fix-a-Form U.S. PAT. NO. 4323608

a. How many milliliters of ampicillin would you give? ____________

____

____

____

47. Order: Tetracycline 1 g/d in two divided doses
Drug available:

SQUIBB

473 mL NDC 0003-0815-50
125 mg per 5 mL
SUMYCIN SYRUP
Tetracycline Oral Suspension USP
125 mg per 5 mL tetracycline hydrochloride equivalent
Caution: Federal law prohibits dispensing without prescription
6505-00-656-1344

SUMYCIN SYRUP
Tetracycline Oral Suspension USP
Bulk Container—Not For Household Dispensing
Each 5 mL (teaspoonful) contains tetracycline equivalent to 125 mg tetracycline hydrochloride, buffered with potassium metaphosphate
Usual pediatric dosage: 10-20 mg per lb of body weight daily in 4 divided doses
See insert
Shake well before using
Keep tightly closed • Protect from light
Store below 30° C (86° F)
Contents should be used within 2 months after bottle is opened since some discoloration may occur
Dispense in tight, light-resistant containers
E. R. Squibb & Sons, Inc.
Princeton, NJ 08540
Made in USA M7048D / 81550

(Convert grams to milligrams.)

a. How many milligrams would the patient receive per dose?

____

b. How many milliliters would the patient receive per dose?

____

48. Order: Potassium chloride 30 mEq, PO, q.d. with food
Drug available: Potassium chloride 20 mEq/15 mL

a. How many milliliters of potassium chloride would the patient receive? ____________

____

49. Order: Docusate sodium (Colace) 200 mg, PO, q.d. per NG tube
Drug available: Colace 50 mg/5 mL

a. How many milliliters of Colace would the patient receive? ____________

50. Order: Ethosuximide (Zarontin) 20 mg/kg/d, PO
Patient weighs 60 kg
Parameter: maximum dose 1.5 g/d
Drug available:

Usual Dosage—*Initially*, 1 or 2 capsules, increased gradually according to the patient's response. The *optimal* dose is 20 mg/kg/day. See package insert for complete prescribing information.
Keep this and all drugs out of the reach of children.
Exp date and lot
0237G161

N 0071-0237-24
Zarontin®
(Ethosuximide Capsules, USP)
250 mg
Caution—Federal law prohibits dispensing without prescription.
100 CAPSULES
PARKE-DAVIS
People Who Care

Dispense in tight container as defined in the USP.
Store at controlled room temperature 15°-30°C (59°-86°F).
6505-00-782-6762
N 3 0071-0237-24 7
US Patent 2,993,835
© 1992, Warner-Lambert Co.
PARKE-DAVIS
Div of Warner-Lambert Co
Morris Plains, NJ 07950 USA

a. How many milligrams would the patient receive per day? ______________________

b. Is this dose within the parameter?________________. Explain:______________

____

c. How many milliliters would the patient receive?______________________

____

51. Order: Amantadine (Symmetrel) 75 mg, PO, b.i.d.
Drug available:

a. How many milliliters would the patient receive per dose? ______________________

NDC 0056-0206-16
NSN 6505-01-130-1945
DUPONT PHARMACEUTICALS
DU PONT **SYMMETREL®**
(amantadine hydrochloride)
Each teaspoonful (5 mL) contains:
Amantadine hydrochloride .... 50 mg
CAUTION: Federal law prohibits dispensing without prescription.
DOSAGE: For dosage and full prescribing information read accompanying product insert.
Dispense in a tight container as defined in the U.S.P.
KEEP TIGHTLY CLOSED
Store at controlled room temperature (59°-86° F, 15°-30° C).
ONE PINT (480 mL) SYRUP

52. Order: Cefadroxil (Duricef) 1 g/d, PO, in two divided doses
Drug available:

OPEN ALONG PERFORATION
Pharmacist: See base label for dispensing directions. Physician leaflet enclosed. Remove before dispensing.
PRINCETON PHARMACEUTICAL PRODUCTS
N 3 0087-0782-41 0
NDC 0087-0782-41
100 ML (WHEN MIXED)
(CEFADROXIL MONOHYDRATE, USP)
FOR ORAL SUSPENSION **250 MG** **5 ML**
NSN 6505-01-153-4351
CAUTION: FEDERAL LAW PROHIBITS DISPENSING WITHOUT PRESCRIPTION

OPEN ALONG PERFORATION
Pharmacist: See base label for dispensing directions. Physician leaflet enclosed. Remove before dispensing.
PRINCETON PHARMACEUTICAL PRODUCTS
N 3 0087-0783-41 7
NDC 0087-0783-41
100 ML (WHEN MIXED)
(CEFADROXIL MONOHYDRATE, USP)
FOR ORAL SUSPENSION **500 MG** **5 ML**
CAUTION: FEDERAL LAW PROHIBITS DISPENSING WITHOUT PRESCRIPTION

a. How many milligrams would the patient receive per dose? ______________________

b. Which bottle of Duricef would you use? ______________________

c. How many milliliters would you give per dose? ______________________

____

53. Order: Cephalexin 0.5 g, PO, q6h
Drug available:

Cephalexin Lederle
for Oral Suspension, USP
250 mg per 5 mL
IMPORTANT -
FOR ORAL USE ONLY.
STORE RECONSTITUTED SUSPENSION IN REFRIGERATOR. DISCARD AFTER 14 DAYS.
Date of Reconstitution

SAMPLE

SHAKE WELL BEFORE USING. KEEP TIGHTLY CLOSED.
Control No.

NDC 0005-3230-46
Cephalexin
for Oral Suspension, USP
250 mg per 5 mL
when reconstituted according to directions.
Usual Children's Dose: 25 to 50 mg per kg per day in four divided doses. For more severe infections, dose may be doubled. See accompanying literature.
WARNING: NOT FOR INJECTION.
Store Dry Powder at Controlled Room Temperature 15-30°C (59-86°F).
CAUTION: Federal law prohibits dispensing without prescription.
100 mL (when reconstituted)
Exp. Date
STANDARD Lederle PRODUCTS

TO THE PHARMACIST
Use this bottle for dispensing. REMOVE THIS PORTION OF LABEL ONLY.
Prepare suspension at time of dispensing. Add to the bottle a TOTAL OF 71 mL of water. For ease in preparation, tap bottle to loosen powder, add the water in 2 portions, shaking well after each addition. The resulting suspension will contain cephalexin monohydrate equivalent to 250 mg cephalexin in each 5 mL (teaspoonful).
Manufactured by
BIOCRAFT LABORATORIES, INC.
Elmwood Park, NJ 07407
for LEDERLE LABORATORIES DIVISION
American Cyanamid Company
Pearl River, NY
10965
24808 BC3
3 0005-3230-46 2
N

a. How many milliliters would the patient receive per dose? (Convert grams to milligrams.)

____

54. Order: Prochlorperazine (Compazine) 7.5 mg, PO, q.i.d.
Drug available:

a. How many milliliters of Compazine would the patient receive per dose?

____

3 0007-3363-44 7
Store at controlled room temperature (59°-86°F).
Dispense in a tight, light-resistant glass bottle.
Each 5 mL (1 teaspoon) contains prochlorperazine, 5 mg, as the edisylate.
Usual Dosage: Children: 5 to 15 mg daily. Adults: 10 to 30 mg daily. See accompanying folder for complete prescribing information.
Important: Use child-resistant closures when dispensing this product unless otherwise directed by physician or requested by purchaser.
Caution: Federal law prohibits dispensing without prescription.
SmithKline Beecham Pharmaceuticals
Philadelphia, PA 19101

5mg/5mL
NDC 0007-3363-44
COMPAZINE®
PROCHLORPERAZINE as the edisylate SYRUP
4 fl oz (118 mL)
SB SmithKline Beecham

55. Order: Chlorpromazine (Thorazine) 25 mg, PO, q.i.d.
Drug available:

a. How many milliliters of Thorazine would the patient receive per dose?

____

____

____

3 0007-5072-44 6
NSN 6505-01-156-1640
Store below 25°C (77°F). Dispense in a tight, light-resistant glass bottle.
Each 5 mL (1 teaspoon) contains chlorpromazine hydrochloride, 10 mg.
Usual Dosage: Children: 10 to 60 mg daily. Adults: 20 to 150 mg daily. See accompanying prescribing information.
Important: Use child-resistant closures when dispensing this product unless otherwise directed by physician or requested by purchaser.
Caution: Federal law prohibits dispensing without prescription.
Manufactured by
SmithKline Beecham Pharmaceuticals
Philadelphia, PA 19101
Marketed by SCIOS NOVA INC.

10mg/5mL
NDC 0007-5072-44
THORAZINE®
CHLORPROMAZINE HCl SYRUP
4 fl oz (118 mL)
SB SmithKline Beecham

56. Order: Theophylline 120 mg, PO, q.d.
Drug available: Theophylline elixir 80 mg/15 mL

a. How many milliliters of theophylline would you give?______________________________

57. Order: Theophylline 5 mg/kg, PO, loading dose
Patient weight: 55 kg
Drug available: Theophylline elixir 50 mg/5 mL

a. How many milliliters of theophylline would you give?____________________

____

58. Order: Amoxicillin/clavulavate (Augmentin) 0.25 g, PO, q8h
Drug available:

AUGMENTIN®

Tear along perforation

NSN 6505-01-340-0847

Directions for mixing:
Tap bottle until all powder flows freely. Add approximately 2/3 of total water for reconstitution (total = 67 mL); shake vigorously to wet powder. Add remaining water; again shake vigorously.
Dosage: See accompanying prescribing information.

Tear along perforation

Keep tightly closed. Shake well before using. Must be refrigerated. Discard after 10 days.

125mg/5mL
NDC 0029-6085-39

AUGMENTIN®
AMOXICILLIN/
CLAVULANATE POTASSIUM
FOR ORAL SUSPENSION
When reconstituted, each 5 mL contains:
AMOXICILLIN, 125 MG,
as the trihydrate
CLAVULANIC ACID, 31.25 MG,
as clavulanate potassium

75mL (when reconstituted)

SB SmithKline Beecham

Use only if inner seal is intact.
Net contents: Equivalent to 1.875 g amoxicillin and 0.469 g clavulanic acid.
Store dry powder at room temperature.
Caution: Federal law prohibits dispensing without prescription.
SmithKline Beecham Pharmaceuticals
Philadelphia, PA 19101

3 0029-6085-39 3

EXP.

LOT

9405804-E

a. How many milliliters of Augmentin would you give per dose?____________________

59. Order: Ampicillin 0.5 g, PO, q8h
Drug available:

a. How many milliliters of ampicillin would you give per dose?

____

____

Control

Exp

Ampicillin 250 mg/5 mL Lederle
Shake well before using. Keep tightly closed.
Notice to patient: Bottle is not full to allow for shaking.
Store in refrigerator. Discard unused portion after 14 days.

NDC 0005-3589-46
Ampicillin
for Oral Suspension, USP
250 mg
ampicillin per 5 mL
1 bottle of powder
(5 g ampicillin anhydrous)
to make 100 mL liquid.
USUAL DOSAGE:
Children - 50 to 100 mg/kg/day in divided doses every 6 to 8 hours, not to exceed adult recommended dosage.
Adults - 250 to 500 mg every 6 hours. Higher doses may be needed for severe infections.
See accompanying circular.
CAUTION: Federal law prohibits dispensing without prescription.
100 mL when mixed

STANDARD Lederle PRODUCTS

Store at room temperature [approx 25°C (77°F)] before reconstitution.
PULL
DIRECTIONS FOR MIXING:
Tap bottle to loosen powder. Slowly add **46 mL** of water all in one portion. Shake vigorously immediately after adding the water.
When reconstituted with 46 mL of water, the bottle will contain 100 mL of suspension, 250 mg ampicillin per 5 mL.

BOTTLE SEALED WITH PRINTED PLASTIC BAND. IF BAND IS BROKEN DO NOT ACCEPT

N 3 0005-3589-46 1

Manufactured by
WYETH LABORATORIES INC.
Philadelphia, PA 19101
for LEDERLE LABORATORIES
A Division of American Cyanamid Company
Pearl River, NY 10965
32103-93 WY3 U3589-46-3

60. Order: Methydopa (Aldomet) 400 mg, PO, b.i.d.
Drug available:

a. How many milliliters would you give per dose?____________________

____

This is a bulk package and not intended for dispensing.
Keep container tightly closed.
Protect from freezing.
Store below 26°C (78°F).
Dispense in a tight, light-resistant container.

MSD NDC 0006-3382-74

473 mL ORAL SUSPENSION
ALDOMET®
(METHYLDOPA, MSD)

250 mg per 5 mL
SHAKE WELL BEFORE USING
CAUTION: Federal (USA) law prohibits dispensing without prescription.
MERCK SHARP & DOHME
DIVISION OF MERCK & CO., INC.
WEST POINT, PA 19486, USA

USUAL DOSAGE: See accompanying circular.
Benzoic acid 0.1% and sodium bisulfite 0.2%, added as preservatives. Alcohol 1%.
473 mL | No. 3382 7592002
Lot

61. Order: Albuterol 4 mg, PO, t.i.d.
Drug available: Albuterol 2 mg/5 mL

   a. How many milliliters of albuterol would you give per dose?____________________

62. Order: Cimetidine (Tagamet) 200 mg, PO, q.i.d. with meals
Drug available:

NDC 0108-5014-01
300mg/5mL
TAGAMET®
CIMETIDINE HCI LIQUID
Single-Dose Unit
Each 5 mL dose contains cimetidine hydrochloride equivalent to cimetidine, 300 mg; alcohol, 2.8%.
Caution: Federal law prohibits dispensing without prescription.
Lot Exp.
SmithKline Beecham Pharmaceuticals
Philadelphia, PA 19101 680409-H
SB SmithKline Beecham

   a. How many milliliters of Tagamet would you give?

____

____

____

63. Order: Potassium chloride (KCl) 15 mEq PO, b.i.d.
Drug available: Potassium chloride 10 mg/15 mL

   a. How many milliliters of potassium chloride would you give per dose? ____________________
   b. How many milliequivalents (mEq) of KCl would the patient receive per day?

____

How many milliliters of KCl per day?____________________

64. Order: Furosemide (Lasix) 30 mg, PO, b.i.d.
Drug available: Furosemide 8 mg/mL

   a. How many milliliters of furosemide would you give?____________________

65. Order: Acyclovir (Zovirax) 400 mg, PO, b.i.d.
Drug available:

1 pint (473 mL)
NDC 0081-0953-96
ZOVIRAX®
(ACYCLOVIR)
Suspension
Each 5 mL (1 teaspoonful) contains acyclovir 200 mg and added as preservatives methylparaben 0.1% and propylparaben 0.02%.
SHAKE WELL BEFORE USING.
For indications, dosage, precautions, etc., see accompanying package insert.
Store at 15° to 25°C (59° to 77°F).
Dispense in tight container as defined in the U.S.P.
CAUTION: Federal law prohibits dispensing without prescription.

   a. How many milliliters of Zovirax would you give per dose?____________________

66. Order: Diphenhydramine 50 mg, PO, q6h
Available: Diphenydramine 12.5 mg/5 mL

   a. How many milliliters of diphenydramine would you give per dose?____________________

67. Order: Lactulose 30 g, PO, t.i.d.
Drug available: Lactulose 10 g/15 mL

    a. How many milliliters of lactulose would you give per dose?________________

68. Order: Guaifenesin 400 mg, PO, q4h × 5 days or until cough subsides
Drug available: Guaifenesin 100 mg/5 mL

    a. How many milliliters of guaifenesin would be given per dose?________________

69. Promethazine 25 mg, PO, b.i.d.
Drug available: Promethazine 6.25 mg/5 mL

    a. How many milliliters of promethazine would you give per dose? ________________

    ____

70. Order: Erythromycin 2 g/d in four divided doses
Drug available: Erythromycin 250 mg/5 mL

    a. Four divided doses would be equivalent to every ________________
    b. How many milligrams (mg) would the patient receive per dose? ________________

    ____

    c. How many milliliters (mL) of erythromycin would be given per dose? ________________

71. Order: Clindamycin 300 mg, PO, q8h
Drug available: Clindamycin 75 mg/mL

    a. How many milliliters of clindamycin would you give per dose?________________

72. Order: Nystatin (Mycostatin) 0.5 million units, q.i.d., swish and swallow
Drug available:

60 mL NDC 0003-0588-60
100,000 units per mL
MYCOSTATIN®
ORAL SUSPENSION
Nystatin Oral Suspension USP
SHAKE WELL BEFORE USING
Caution: Federal law prohibits dispensing without prescription
APOTHECON®
A BRISTOL-MYERS SQUIBB COMPANY

Before Dispensing Replace Cap With Safety Cap Dropper
Each mL contains 100,000 USP Nystatin Units in a vehicle containing 50% sucrose. Not more than 1% alcohol by volume. Usual dosage: See insert
Store at room temperature; avoid freezing
APOTHECON®
A Bristol-Myers Squibb Company
Princeton, NJ 08540 USA
P1874-00

    a. How many milliliters of Mycostatin per dose would you pour?________________

INJECTABLES

73. Order: Heparin 2,500 units, S.C., q6h
    Drugs available:

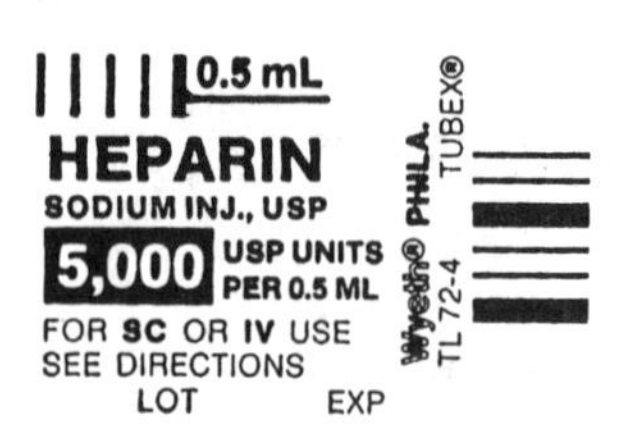

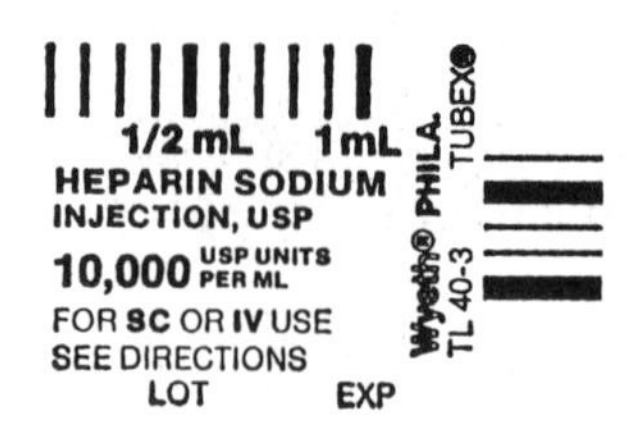

   a. Which cartridge of heparin would you use?____________________
   b. How many milliliters of heparin would you give per dose?____________________

74. Order: Heparin U 8,000, S.C., q6h
    Drugs available:

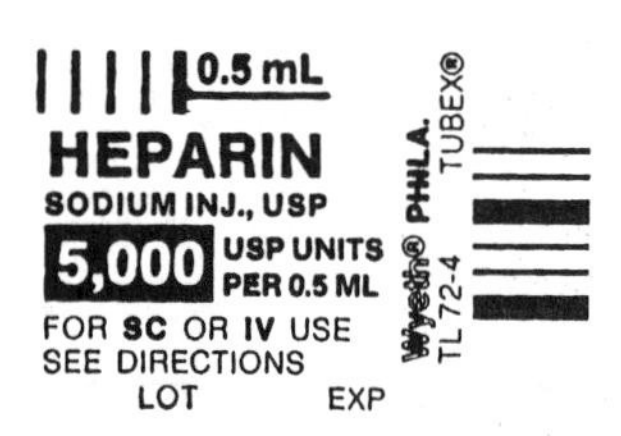

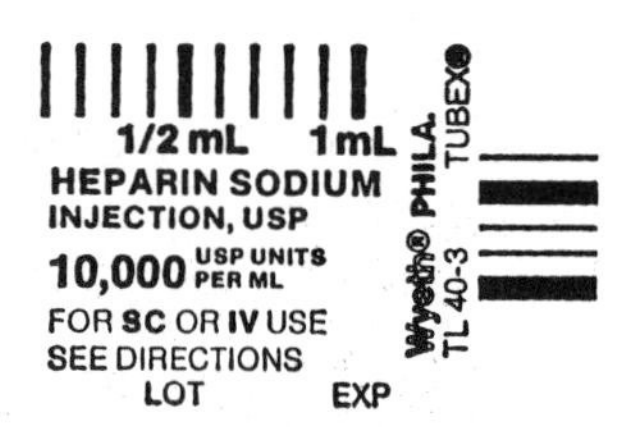

   a. Which cartridge of heparin would you use?____________________
   b. How many milliliters of heparin would you give per dose?____________________

75. Order: Heparin U 7,500, S.C. q6h
    Drug available: Heparin U 10,000/mL, multi-vial

   a. Which type of syringe would you use?____________________
   b. What is the method of injection?____________________
   c. How many milliliters of heparin would you give per dose?____________________

76. Order: Regular insulin 12 units, S.C., stat
    Drug available: Regular insulin U 100/mL
    Insulin syringe: U 100/mL

   a. Indicate on the insulin syringe (below) the amount of insulin you would withdraw.

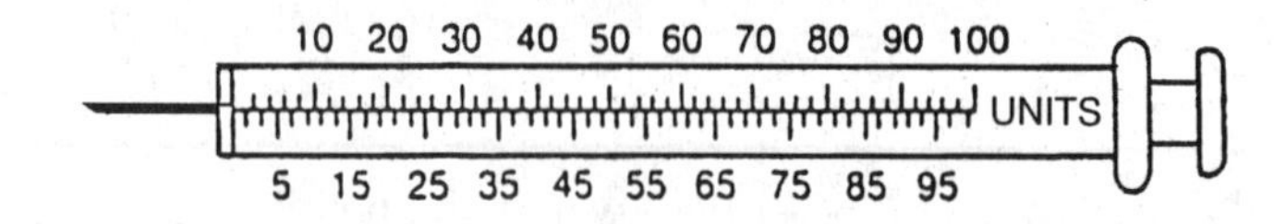

77. Order: NPH insulin U 25, S.C., q.d. (7 a.m.)
Drug available:

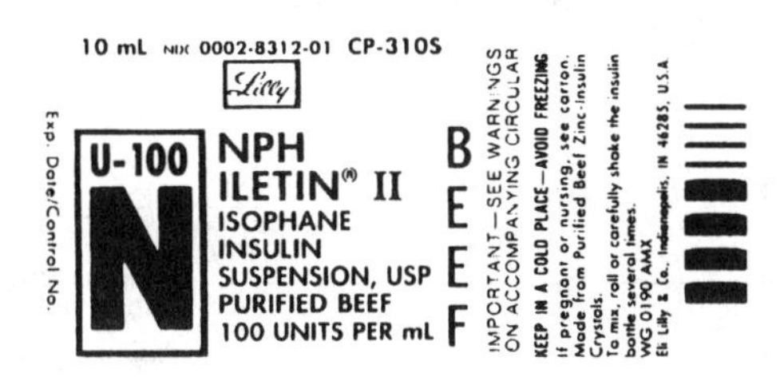

a. Indicate on the insulin syringe the amount of insulin you would withdraw.

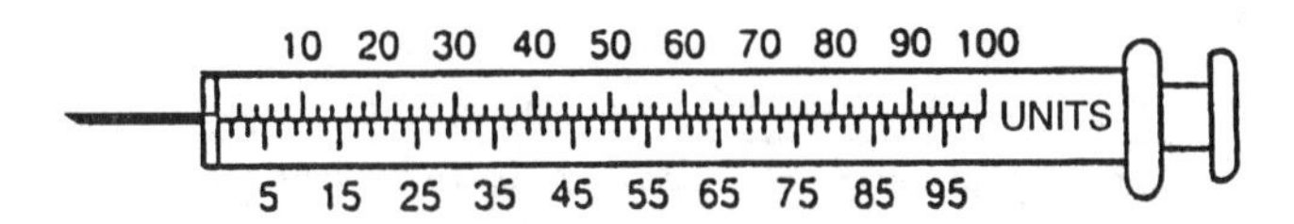

78. Order: Lente insulin U 42, S.C., q.d.
Drug available: Lente Insulin U 100/mL and insulin syringe

a. Indicate on the insulin syringe the amount of insulin you would withdraw.

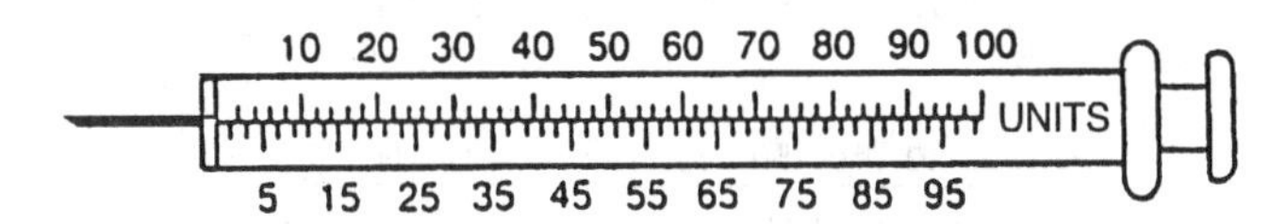

79. Order: Regular insulin 5 units and Lente insulin 45 units, qd
Drugs available:

a. Explain the method for mixing the two insulins. ______________________________

____

____

b. Mark on the insulin syringe how much regular insulin and how much Lente insulin you would withdraw.

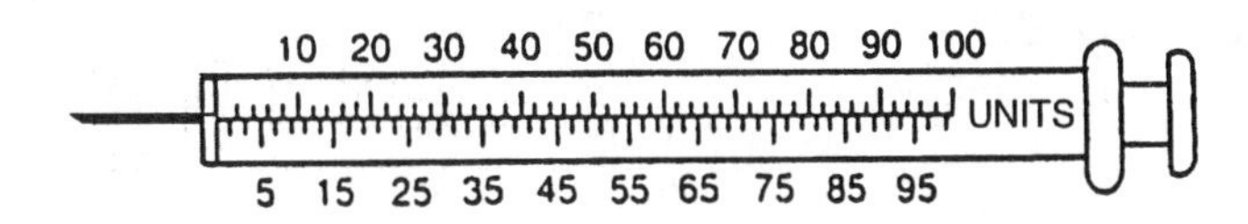

80. Order: Humulin Regular insulin 6 units and Humulin NPH insulin 40 units, S.C., q.d.
Drugs available: Humulin Regular insulin U 100/mL and Humulin NPH insulin U 100/mL
Insulin syringe: U 100/mL

a. Explain the method for mixing the two insulins:______________________________

____

____

b. Mark on the insulin syringe how much regular insulin and how much NPH insulin you would withdraw.

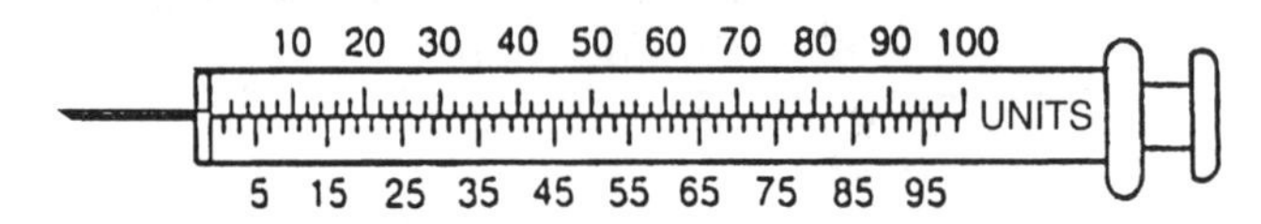

81. Order: Regular insulin 10 units and NPH insulin 36 units
Drugs available: Regular insulin U 100/mL and NPH insulin U 100/mL

a. Mark on the insulin syringe how much regular insulin and how much NPH insulin you would withdraw.

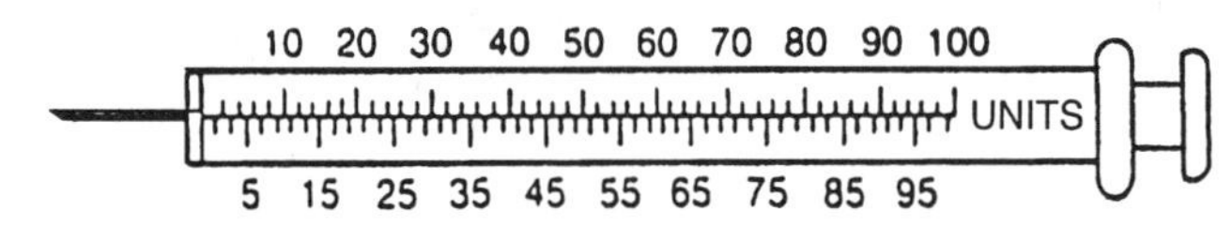

82. Order: Narcan 0.2 mg, IM, stat
Drug available:

NDC 0590-0358-01
NARCAN®
(naloxone HCl injection, USP)
0.4 mg/mL
1 mL AMPUL
FOR IM, SC OR IV USE
DuPont Pharma
Manati, Puerto Rico 00674
Lot:
Exp:

a. How many milliliters of Narcan would you give?

____

b. Can you refrigerate the unused portion of the ampul of Narcan?

____

83. Order: Morphine gr 1/6, IM, stat
Drug available:

(Convert grains to milligrams; use conversion table if needed.)

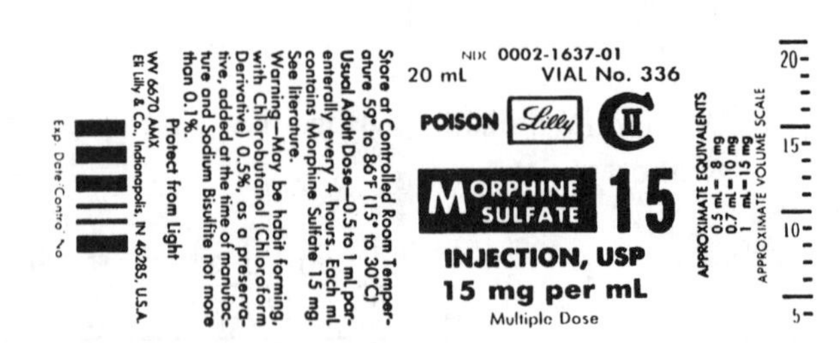

a. How many milliliters of morphine would you give?______________________________

____

84. Order: Vitamin B_{12} 500 mcg (μg), IM, 2 × per week
Drugs available:

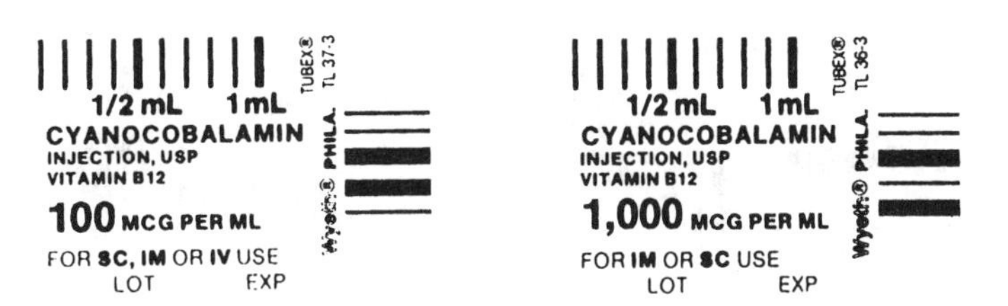

a. Which cyanocobalamin cartridge would you use? ______________________________

____

b. How many milliliters of cyanocobalamin would you give? ______________________________

85. Order: Hydroxyzine (Vistaril) 75 mg, deep IM, stat
Drug available: Vistaril 100 mg/2 mL

a. How many milliliters of hydroxyzine would the patient receive? ______________________________

____

86. Order: Meperidine 30 mg, and atropine SO_4 0.3 mg, IM
Drugs available:

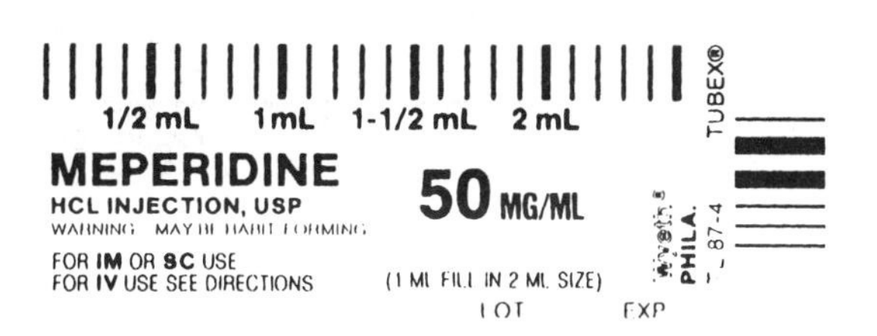

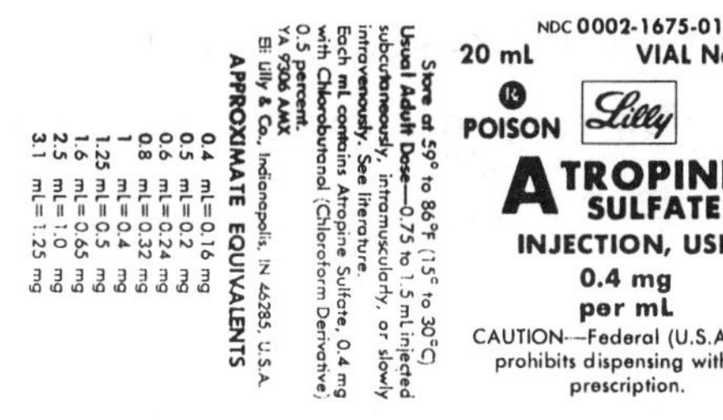

a. How many milliliters of meperidine would you give? ______________________________

How many milliliters of atropine? ______________________________

b. Explain how the two drugs can be mixed in the cartridge: ______________________________

____

____

87. Order: Meperidine 70 mg and Atropine SO_4 gr 1/100, IM, stat
Drugs available:

1/2 mL 1 mL 1-1/2 mL 2 mL
MEPERIDINE 100 MG/ML
HCL INJECTION, USP
WARNING—MAY BE HABIT-FORMING
FOR IM OR SC USE
FOR IV USE SEE DIRECTIONS
(1 ML FILL IN 2 ML SIZE)
LOT EXP

NDC 0002-1675-01
20 mL VIAL No. 419
POISON
ATROPINE SULFATE
INJECTION, USP
0.4 mg per mL
CAUTION—Federal (U.S.A.) law prohibits dispensing without prescription.

(Convert grains to milligrams.)

a. How many milliliters of meperidine would you give?__________

How many milliliters of atropine? __________

b. Explain how the two drugs can be mixed in the cartridge: __________

88. Order: Kanamycin (Kantrex) 400 mg, IM, q8h
Patient's weight: 90 kg
Dose parameter: 15 mg/kg/d in 2 or 3 divided doses
Drug available:

NDC 0015-3503-20
NSN 6505-00-660-1676
EQUIVALENT TO
1 gram KANAMYCIN per 3 mL
KANTREX
Kanamycin Sulfate Injection, USP
FOR I.M. OR I.V. USE
CAUTION: Federal law prohibits dispensing without prescription.

a. What is the dose parameter according to the patient's weight?__________

b. How many milligrams would the patient receive per day? __________

c. Is the drug dose within safe dose parameters?__________

d. How many milliliters of Kantrex would you give per dose?__________

89. Order: Amikacin 500 mg, IM, q12h
Patient's weight: 149 pounds
Dose parameter: 15 mg/kg/d in 2 or 3 divided doses
Drug available:

NDC 0015-3023-20
Amikin
AMIKACIN SULFATE INJECTION For I.M. or I.V. Use
EQUIVALENT TO
1 gram AMIKACIN per 4 mL
CAUTION: Federal law prohibits dispensing without prescription.
©Bristol-Myers Company

a. How many kilograms does the patient weigh?__________

b. What is the dose parameter according to the patient's weight?__________

c. Is the drug dose within safe dose parameters?__________

d. How many milliliters of amikacin would you give per dose? __________

90. Order: Chlorpromazine HCl (Thorazine) 20 mg, deep IM, t.i.d.
Drug available:

NSN 6505-01-156-1981
Dilute before I.V. use. Store below 86°F.
Do not freeze. Protect from light.
Each mL contains, in aqueous solution, chlorpromazine hydrochloride, 25 mg, ascorbic acid, 2 mg, sodium bisulfite, 1 mg, sodium sulfite, 1 mg, sodium chloride, 1 mg. Contains benzyl alcohol, 2%, as preservative.
See accompanying prescribing information.
For deep I.M. injection.
Caution: Federal law prohibits dispensing without prescription.
Manufactured by **SmithKline Beecham Pharmaceuticals**, Philadelphia, PA 19101
Marketed by SCIOS NOVA INC
LOT EXP.
693897-AE

25mg/mL
NDC 0007-5062-01
THORAZINE®
CHLORPROMAZINE HCl INJECTION
10 mL Multi-Dose Vial
SB SmithKline Beecham

a. How many milliliters of Thorazine would you give per dose? ____________

91. Order: Cimetidine HCl (Tagamet) 100 mg, IM, q.i.d.
Drug available:

Store between 15° and 30°C (59° and 86°F).
Do not refrigerate.
Each 2 mL Single-dose ADD-Vantage®* Vial contains, in aqueous solution, cimetidine hydrochloride equivalent to cimetidine, 300 mg; phenol, 10 mg.
Dosage: See accompanying prescribing information and instructions for using vial.
Caution: Federal law prohibits dispensing without prescription.
SmithKline Beecham Pharmaceuticals
Philadelphia, PA 19101
LOT EXP.
670630-H

300mg/2mL
NDC 0108-5031-01
TAGAMET®
CIMETIDINE HCl INJECTION
SB SmithKline Beecham

2 mL Single-Dose ADD-Vantage®* Vial
For I.V. Infusion
*ADD-Vantage® is a trademark of Abbott Laboratories.
Tagamet® Injection

a. How many milliliters of Tagamet would the patient receive per dose? ____________

92. Order: Prochlorperazine (Compazine) 7.5 mg, deep IM, stat
Drug available:

Store below 86°F. Do not freeze.
Protect from light. Discard if markedly discolored.
Each mL contains, in aqueous solution, prochlorperazine, 5 mg, as the edisylate; sodium biphosphate, 5 mg; sodium tartrate, 12 mg; sodium saccharin, 0.9 mg; benzyl alcohol, 0.75%, as preservative.
Dosage: For deep I.M. or I.V. injection.
See accompanying prescribing information.
Caution: Federal law prohibits dispensing without prescription.
SmithKline Beecham Pharmaceuticals
Philadelphia, PA 19101
LOT EXP.
693793-AD

10mL Multi-Dose Vial
5mg/mL
NDC 0007-3343-01
COMPAZINE®
PROCHLORPERAZINE as the edisylate INJECTION
SB SmithKline Beecham

a. How many milliliters of Compazine would the patient receive? ____________

93. Order: Compazine 2.5 mg, deep IM, q4h
Drug available:

Store below 86°F. Do not freeze.
Protect from light. Discard if markedly discolored.
Each mL contains, in aqueous solution, prochlorperazine, 5 mg, as the edisylate; sodium biphosphate, 5 mg; sodium tartrate, 12 mg; sodium saccharin, 0.9 mg; benzyl alcohol, 0.75%, as preservative.
Dosage: For deep I.M. or I.V. injection.
See accompanying prescribing information.
Caution: Federal law prohibits dispensing without prescription.
SmithKline Beecham Pharmaceuticals
Philadelphia, PA 19101
LOT EXP.
693793-AD

10mL Multi-Dose Vial
5mg/mL
NDC 0007-3343-01
COMPAZINE®
PROCHLORPERAZINE as the edisylate INJECTION
SB SmithKline Beecham

a. How many milliliters of Compazine would you give per dose? ____________

94. Order: Haloperidol (Haldol) decanoate 100 mg, IM, q 4 weeks.
Drug available:

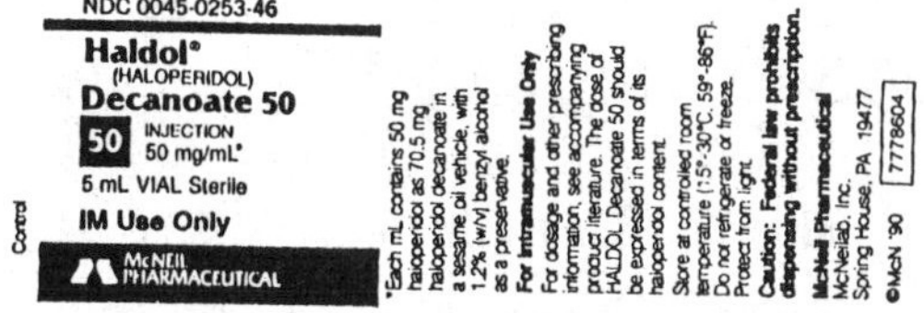

a. How many milliliters of Haldol would you give? ____________

95. Order: Ticarcillin (Ticar) 500 mg, IM, q6h
Drug available:

3 0029-6550-22 8
equivalent to
1gram ticarcillin
NDC 0029 6550 22
TICAR®
STERILE TICARCILLIN DISODIUM INJECTION
For I.M. or I.V. Use
SB SmithKline Beecham

NSN 6505-01-046-6794
Store dry powder at room temperature or below.
I.V. Use: Add at least 4 ml Sterile Water for Injection, USP, when dissolved, dilute further to desired volume with water or an appropriate I.V. solution.
I.M. Use: Add 2 ml Sterile Water for Injection, USP, or 1% Lidocaine HCl solution (without epinephrine) and **use promptly.** Each 2.6 ml of solution will then contain 1 gram of ticarcillin.
Dosage: See accompanying prescribing information, including dosage and stability in I.V. solutions.
Caution: Federal law prohibits dispensing without prescription
SmithKline Beecham Pharmaceuticals
Philadelphia, PA 19101
9516400-G
LOT
EXP.

a. How many milliliters of diluent would you add to the powdered ticarcillin? (see drug label) ____________

b. After reconstitution, you would have how many milliliters of ticarcillin solution?
____

c. How many milliliters of Ticar would the patient receive per dose? ____________
____

96. Order: Cefazolin (Ancef) 0.25 g, IM, q6h
Drug available:

EXP.
equivalent to
500mg cefazolin
NDC 0007-3131-16
ANCEF®
STERILE CEFAZOLIN SODIUM (LYOPHILIZED)
LOT
25 Vials for Intramuscular or Intravenous Use

NSN 6505-01-308-5346
Before reconstitution protect from light and store at controlled room temperature (15° to 30°C; 59° to 86°F).
Usual Adult Dosage: 250 mg to 1 gram every 6 to 8 hours. See accompanying prescribing information.
For I.M. administration add 2.0 mL of Sterile Water for Injection. SHAKE WELL. Withdraw entire contents. Provides an approximate volume of 2.2 mL (225 mg/mL). For I.V. administration see accompanying prescribing information.
Reconstituted *Ancef* is stable for 24 hours at room temperature or for 10 days if refrigerated (5°C or 41°F).
SmithKline Beecham Pharmaceuticals
Philadelphia, PA 19101
694117-A
K3131-16
3 0007-3131-16 9

a. How many milliliters of diluent would you add? (see drug label) ____________

b. After reconstitution, you would have how many milliliters of cefazolin solution?
____

c. How many milliliters of Ancef would you administer per dose? ____________
____

97. Order: Oxacillin 250 mg, IM, q6h
Drug available:

NDC 0015-7981-20
EQUIVALENT TO
1 gram OXACILLIN
OXACILLIN SODIUM FOR INJECTION, USP
Buffered—For IM or IV Use
CAUTION: Federal law prohibits dispensing without prescription.
APOTHECON® A BRISTOL-MYERS SQUIBB COMPANY

This vial contains oxacillin sodium monohydrate equivalent to 1 gram oxacillin and 20 mg dibasic sodium phosphate. Add 5.7 mL Sterile Water for Injection, USP. • Each 1.5 mL contains 250 mg oxacillin. • Usual Dosage: Adults—250 mg to 500 mg intramuscularly every 4 to 6 hours. See circular for intravenous use. READ ACCOMPANYING CIRCULAR. Discard solution after 3 days at room temperature or 7 days under refrigeration.
APOTHECON® A Bristol-Myers Squibb Company Princeton, NJ 08540 USA 798120DRL-2
Cont:
Exp. Date:

a. How much diluent would you add to the Oxacillin vial? ____________

b. How many milliliters of Oxacillin would you give per dose? ____________

98. Order: Nafcillin 0.5 g, IM, q6h
Drug available:

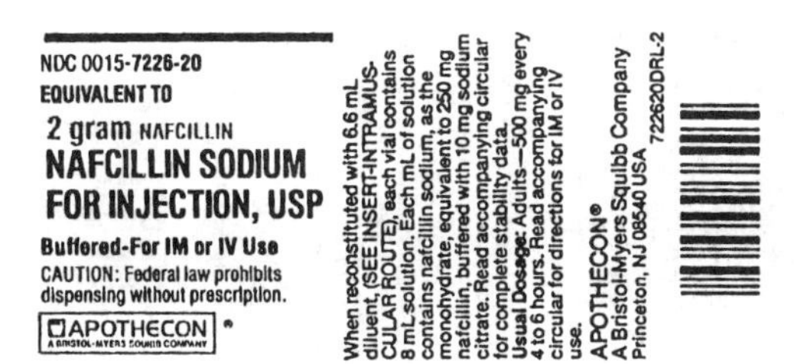

a. How many milliliters of diluent would you add? (see drug label)________________

b. After reconstitution, you would have how many milliliters of cefazolin solution?

____

c. How many milliliters of nafcillin would you give per dose?________________

____

99. Order: Tagamet 0.2 g, IM, q6h.
Drug available:

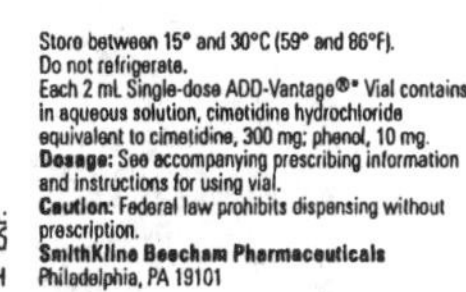

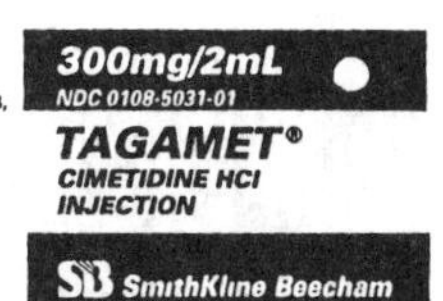

a. How many milliliters of Tagamet would you give per dose?________________

____

100. Order: Dyphylline 0.5 g, IM, q6h.
Drug available: Dyphylline 250 mg/mL

a. How many milliliters of Dyphylline would you give per dose?________________

101. Order: Clindamycin 250 mg, IM, q6h
Drug available:

a. How many milliliters of clindamycin would you give per dose? ________________

102. Order: Ampicillin 500 mg, IM, q6h.
Drug available:

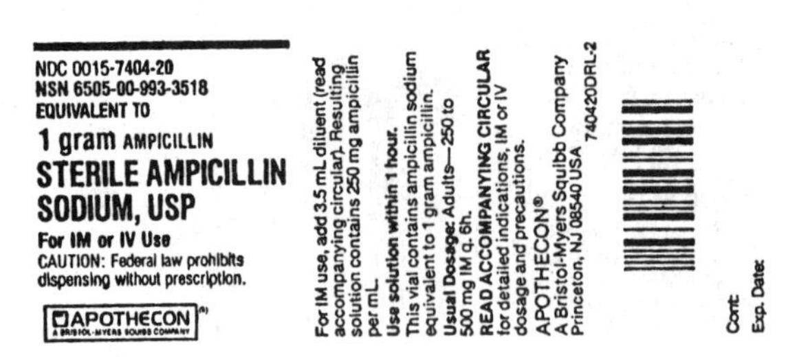

a. How many milliliters of diluent would you add to the ampicillin vial? (see drug label) ______

b. After reconstitution, you would have how many milliliters of ampicillin solution? ______

c. How many milliliters of ampicillin would you give per dose? ______

______

103. Order: Cefamandole (Mandol) 750 mg, IM, q8h
Drug available:

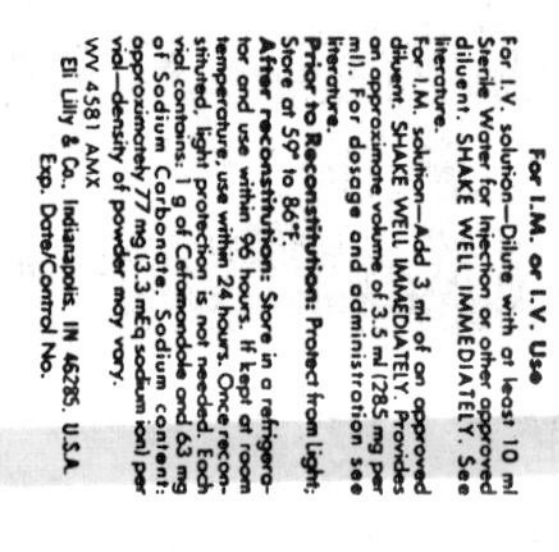

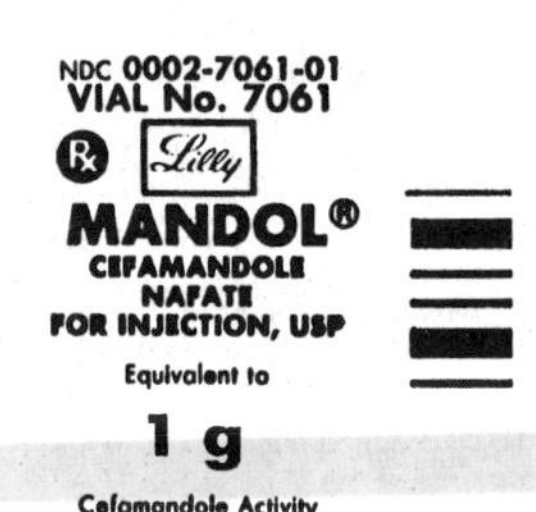

a. How many milliliters of diluent would you add to the vial? (see drug label) ______

b. After reconstitution, you would have how many milliliters of Mandol solution? ______

c. How many milliliters of Mandol would you give per dose? ______

______

______

104. Order: Morphine 3 mg, IM, q4-6h, PRN × 3 days
Drug available:

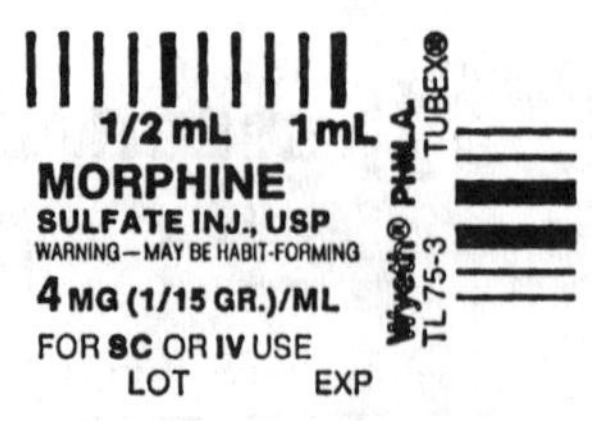

a. How many milliliters of morphine would you give per dose? ______

105. Order: Cephapirin (Cefadyl) 600 mg, IM, q8h
Drug available:

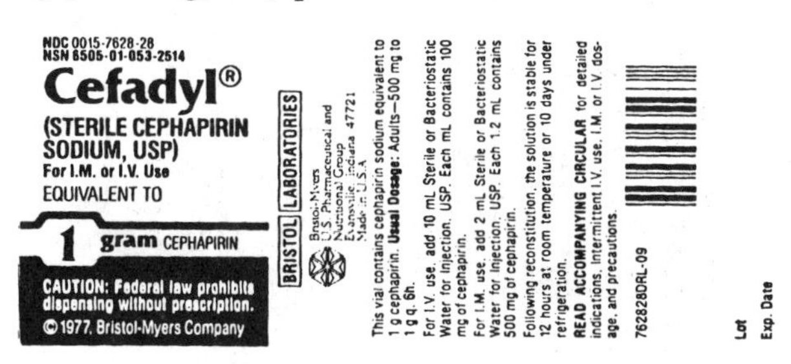

a. How many milliliters of diluent would you add to the vial? (see drug label) ________

b. How many milliliters of reconstituted solution are equivalent to 1 g of Cefadyl solution?

____

c. How many milliliters of Cefadyl would you give per dose? ________

____

106. Order: Dexamethasone (Decadron) 3 mg, IM, stat
Drug available:

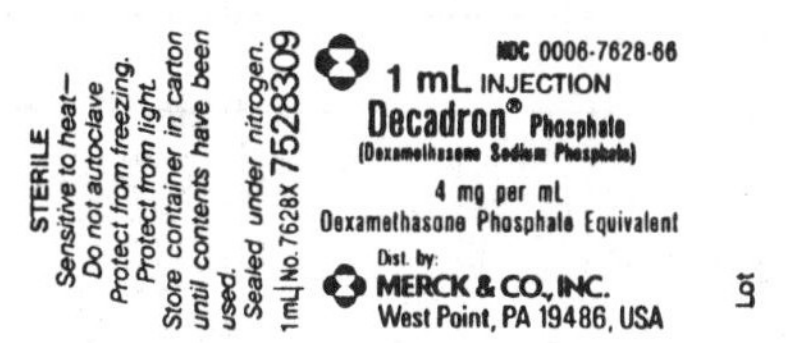

a. How many milliliters of Decadron would you give per dose? ________

107. Atropine SO_4 gr 1/100, IM, stat
Drug available:

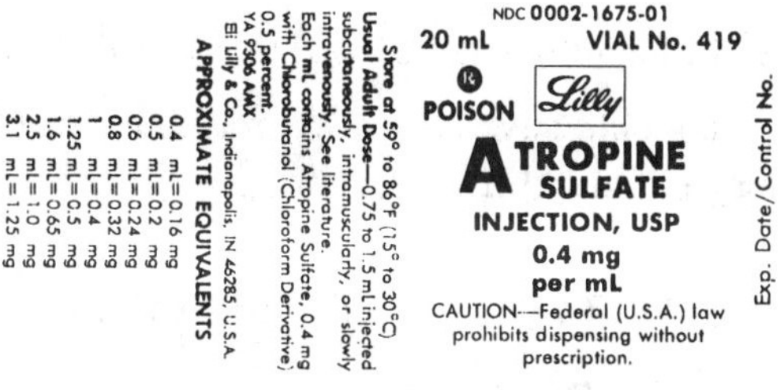

a. How many milliliters of Atropine would you give? ________

____

108. Order: Moxalactam (Moxam) 3 g, IM, q8h
Drug available: Moxam 10 g vial; reconstituted: add 8 mL diluent, which equals 10 mL of reconstituted solution

a. How many milliliters of reconstituted solution are equivalent to 3 g of Moxam?

____

____

109. Order: Tobramycin SO_4 (Nebcin) 3–5 mg/kg/d, IM, in three divided doses
Patient weight: 64 kg
Drug available:

NDC 0002-1499-01
2 mL VIAL No. 781
NEBCIN®
TOBRAMYCIN SULFATE INJECTION USP
80 mg per 2 mL
Multiple Dose
For I.M. or I.V. Use
Must dilute for I.V. use.

a. How many milligrams of tobramycin would the patient receive per day? ________________

____

____

b. How many milligrams of tobramycin would you give per dose? ________________

c. How many milliliters of tobramycin would you give per dose? ________________

____

____

110. Order: Aztreonam (Azactam) 500 mg, IM, q8h
Drug available: Azactam 1 g vial; reconstituted: add 3 mL of diluent, which equals 3.5 mL of reconstituted solution

a. How many milliliters of reconstituted solution would you give per dose? ________________

111. Order: Meperidine (Demerol) 35 mg, IM, q4-6h, prn
Drug available:

1/2 mL 1 mL 1-1/2 mL 2 mL
MEPERIDINE
HCL INJECTION, USP
50 MG/ML
FOR IM OR SC USE
FOR IV USE SEE DIRECTIONS
(1 ML FILL IN 2 ML SIZE)
LOT EXP
TUBEX®

a. How many milliliters of Demerol would you give per dose? ________________

PEDIATRICS

112. Child has a respiratory infection.
Order: Cefixime (Suprax) 80 mg, PO, q12h
Child's weight: 44 lbs, or 20 kg
Pediatric dose parameter: 8 mg/kg/d in a single dose or two divided doses
Drug available: Suprax 100 mg/5 mL oral suspension

a. Is the dose within safe dose parameters?________________

Explain:________________

b. How many milliliters of Suprax would you give per dose?________________

113. Small child has an ear infection.
Order: Amoxicillin (Amoxil) 75 mg, PO, q8h
Child's weight: 70 lbs, or 32 kg
Pediatric dose parameter: 20–40 mg/kg/d in three divided doses
Drug available:

AMOXIL®

Tear along perforation
Directions for mixing: Tap bottle until all powder flows freely. Add approximately 1/3 total amount of water for reconstitution (total=62 mL); shake vigorously to wet powder. Add remaining water; again shake vigorously. Each 5 mL (1 teaspoonful) will contain amoxicillin trihydrate equivalent to 125 mg amoxicillin.
Usual Adult Dosage: 250 to 500 mg every 8 hours.
Usual Child Dosage: 20 to 40 mg/kg/day in divided doses every 8 hours, depending on age, weight and infection severity. See accompanying prescribing information.
Tear along perforation

Keep tightly closed.
Shake well before using.
Refrigeration preferable but not required.
Discard suspension after 14 days.

125mg/5mL
NDC 0029-6008-21

AMOXIL®
AMOXICILLIN
FOR ORAL
SUSPENSION

80mL (when reconstituted)

SB SmithKline Beecham

NSN 6505-01-153-3415
Net contents: Equivalent to 2.0 grams amoxicillin.
Store dry powder at room temperature.
Caution: Federal law prohibits dispensing without prescription.
SmithKline Beecham Pharmaceuticals
Philadelphia, PA 19101

3 0029-6008-21 7

EXP.

LOT

9405773-B

a. How many milligrams of amoxicillin would the child receive per day?________________

a. Is the dose within safe dose parameters?________________

Explain:________________

b. How many milliliters of Amoxil would you give per dose? ________________

114. Child has a staph infection.
Order: Clindamycin 150 mg, PO, q6h
Child's weight: 40 lbs or 18 kg
Pediatric dose parameter: 8–25 mg/kg/d in three or four divided doses
Drug available: Clindamycin 75 mg/mL oral solution

a. How many milligrams of clindamycin would the child receive per day?________________

b. Is the dose within safe dose parameters?________________

Explain:________________

c. How many milliliters of clindamycin would you give per dose?________________

____

115. Child has otitis media.
    Order: Cefaclor (Ceclor) 0.25 g, PO, q8h
    Child's weight: 66 lbs
    Pediatric dose parameter: 20–40 mg/kg/d in three divided doses; maximum: 1 g/d
    Drug available: Ceclor 250 mg/5 mL

    a. How many kilograms does the child weigh?____________________

    b. How many milligrams would the child receive per day?____________________

    c. Is the dose within safe dose parameters? ____________________

       Explain:____________________

    d. How many milliliters of Ceclor would you give per dose? ____________________

116. Child has otitis media.
    Order: Cefaclor (Ceclor) 100 mg, PO, q8h
    Child's weight: 30 lbs
    Pediatric dose parameter: 20–40 mg/kg/d in three divided doses
    Drug available: Ceclor 125 mg/5 mL

    a. How many kilograms does the child weigh?____________________

    b. How many milligrams would the child receive per day?____________________

    c. Is the dose within safe dose parameters?____________________

    d. How many milliliters of Ceclor would you give per dose?____________________

117. Child has a strept infection
    Order: Penicillin V 100 mg, PO, q6h
    Child's weight: 14 kg
    Pediatric dose parameter: 15–50 mg/kg/d in three to six divided doses
    Drug available:

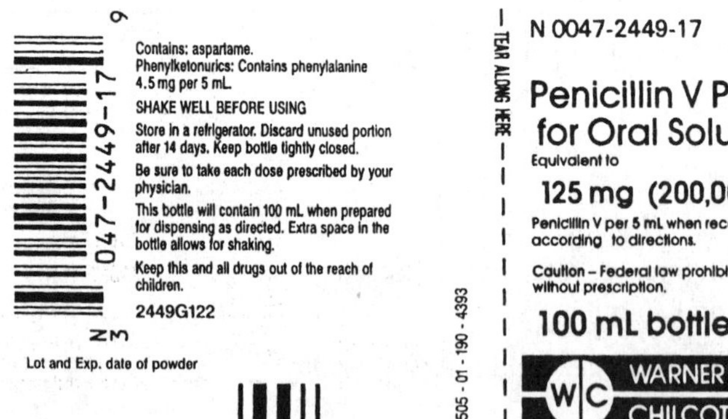

N 0047-2449-17
RECONSTITUTE w/ 63 mL WATER
Penicillin V Potassium for Oral Solution, USP
Equivalent to
125 mg (200,000 units)
Penicillin V per 5 mL when reconstituted according to directions.
Caution – Federal law prohibits dispensing without prescription.
100 mL bottle
WARNER CHILCOTT

Dispensing Directions – Prepare solution at time of dispensing. Shake bottle to loosen powder or break up the cake that may have formed. Add a total of 63 mL water in 2 portions and shake well after each. This provides 100 mL of solution.
Usual Dosage – 125 or 250 mg every 6 to 8 hours. See package insert.
Store below 30°C (86°F).
Protect from moisture.
Manufactured for: WARNER CHILCOTT LABS
Div of Warner-Lambert Co ©1990
Morris Plains NJ 07950 USA
By: Clonmel Chemicals Co. Ltd.
Clonmel, Republic of Ireland
2449J012

    a. How many milligrams would the child receive per day?____________________

    b. Is the dose within safe dose parameters? ____________________

    c. How many milliliters would you give the child per dose?____________________

118. Child has seizures.
Order: Phenytoin (Dilantin) 50 mg, PO, b.i.d.
Child's weight: 16 kg
Pediatric dose parameter: 3–8 mg/kg/d
Drugs available: Dilantin 30 mg/5 mL and Dilantin 100 mg/5 mL

a. How many milligrams would the child receive per day?____________________

b. What is the dose parameter according to the child's weight?____________________

____

c. Is the dose within safe dose parameters? ____________________

d. How many milliliters would you give the child per dose?____________________

119. Child has seizures.
Order: Phenytoin (Dilantin) 125 mg, PO, q.d.
Child's height and weight: 36 in., 45 lbs
Pediatric dose parameter: 250 mg/m^2/d
Drugs available: Dilantin 30 mg/5 mL and 100 mg/5 mL

a. What is the child's body surface area (BSA) in meters squared (m^2)?
(Use nomogram to determine BSA.) ____________________

b. What is the dose parameter according to the child's BSA?____________________

____

c. Is the dose within safe dose parameters? ____________________

d. Which bottle of Dilantin would you use?____________________

e. How many milliliters would you give the child per day?____________________

120. Child has a respiratory infection.
Order: Cefadroxil (Duricef) 400 mg, PO, q12h
Child's weight: 32 kg
Pediatric dose parameter: 30 mg/kg/d in two divided doses
Drugs available:

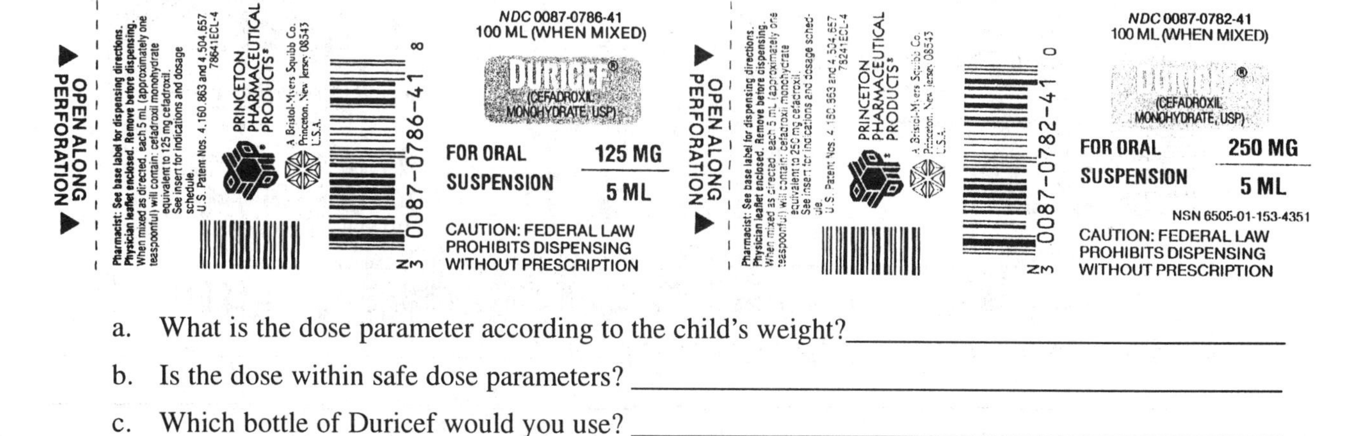

a. What is the dose parameter according to the child's weight?____________________

b. Is the dose within safe dose parameters? ____________________

c. Which bottle of Duricef would you use? ____________________

d. How many milliliters would you give the child per dose?____________________

____

121. Child has a systemic infection.
Order: Ampicillin 150 mg, PO, q6h
Child's weight: 12 kg
Pediatric dose parameter: 50–100 mg/kg/d in four divided doses
Drug available:

NDC 0005-3589-46
Ampicillin
for Oral Suspension, USP
250 mg
ampicillin per 5 mL
1 bottle of powder (5 g ampicillin anhydrous) to make 100 mL liquid.
USUAL DOSAGE:
Children - 50 to 100 mg/kg/day in divided doses every 6 to 8 hours, not to exceed adult recommended dosage.
Adults - 250 to 500 mg every 6 hours. Higher doses may be needed for severe infections.
See accompanying circular.
CAUTION: Federal law prohibits dispensing without prescription.
100 mL when mixed
Store at room temperature [approx 25°C (77°F)] before reconstitution.
DIRECTIONS FOR MIXING:
Tap bottle to loosen powder. Slowly add 46 mL of water all in one portion. Shake vigorously immediately after adding the water.
When reconstituted with 46 mL of water, the bottle will contain 100 mL of suspension, 250 mg ampicillin per 5 mL.
Manufactured by WYETH LABORATORIES INC. Philadelphia, PA 19101 for LEDERLE LABORATORIES A Division of American Cyanamid Company Pearl River, NY 10965
32103-93 WY3 U3589-46-3
Ampicillin 250 mg/5 mL Lederle
Shake well before using. Keep tightly closed.
Notice to patient: Bottle is not full to allow for shaking.
Store in refrigerator. Discard unused portion after 14 days.

a. How many milligrams of ampicillin would the child receive per day?________________

b. Is the dose within safe dose parameters? ________________

Explain:________________

c. How many milliliters would you give the child per dose?________________

122. Child (12 year old) has hypertension.
Order: Methydopa (Aldomet) 250 mg, PO, b.i.d.
Child's weight: 110 lbs
Pediatric dose parameter: 10 mg/kg/d in two to four divided doses
Drug available:

MSD NDC 0006-3382-74
473 mL ORAL SUSPENSION
ALDOMET®
(METHYLDOPA, MSD)
250 mg per 5 mL
SHAKE WELL BEFORE USING
CAUTION: Federal (USA) law prohibits dispensing without prescription.
MERCK SHARP & DOHME DIVISION OF MERCK & CO., INC. WEST POINT, PA 19486, USA
This is a bulk package and not intended for dispensing.
Keep container tightly closed.
Protect from freezing.
Store below 26°C (78°F).
Dispense in a tight, light-resistant container.
USUAL DOSAGE: See accompanying circular.
Benzoic acid 0.1% and sodium bisulfite 0.2%, added as preservatives. Alcohol 1%.
473 mL | No. 3382 7592002
Lot

a. How many kilograms does the child weigh?________________

b. How many milligrams would the child receive per day? ________________

c. Is the dose within safe dose parameters? ________________

Explain:________________

d. How many milliliters of Aldomet would you give the child per dose? ________________

123. Child has pain.
Order: Codeine gr 1/4, PO, q4-6h, prn
Child's height and weight: 42 in., 50 lbs
Pediatric dose parameter: 100 mg/m^2/d
Drug available: Codeine 15 mg tablets

a. What is the child's body surface area (BSA, m^2)? (Use nomogram.)

____

b. What is the dose parameter according to the child's BSA? (Convert grains to milligrams.)

____

____

c. Is the dose within safe dose parameters? ______________

Explain: ______________________________

d. How many tablet(s) would you give the child per dose? ______________

124. Infant (2 month old) has congestive heart failure.
Order: Digoxin (Lanoxin); initially: 0.1 mg, PO, q8h × 3 doses
Child's weight: 8 lbs
Pediatric dose: Child 1 month to 2 years old: 0.035–0.060 mg/kg/d in three divided doses for 24 hr
Drug available:

60 mL NDC 0081-0264-27
LANOXIN®
(DIGOXIN)
ELIXIR
PEDIATRIC
Each ml contains
50 μg (0.05 mg)
PLEASANTLY FLAVORED
Wellcome
BURROUGHS WELLCOME CO.
RESEARCH TRIANGLE PARK, NC 27709
Made in U.S.A. 542402
Alcohol 10%, Methylparaben 0.1% (added as a preservative)
For indications, dosage, precautions, etc., see accompanying package insert.
Store at 15° to 25°C (59° to 77°F) and protect from light.
CAUTION: Federal law prohibits dispensing without prescription.

a. What is the dose parameter according to the infant's weight?

____

____

b. Is the dose within safe dose parameters? ______________

Explain: ______________________________

c. How many milliliters would you give the child per dose?

____

125. Infant (2 month old) has congestive heart failure.
Order: Digoxin (Lanoxin) Maint dose: 0.05 mg, PO, q12h
Child's weight: 8 lbs
Pediatric dose: Child 1 month to 2 years old: 0.01–0.02 mg/kg/d in two divided doses
Drug available:

60 mL NDC 0081-0264-27
LANOXIN®
(DIGOXIN)
ELIXIR
PEDIATRIC
Each ml contains
50 μg (0.05 mg)
PLEASANTLY FLAVORED
Wellcome
BURROUGHS WELLCOME CO.
RESEARCH TRIANGLE PARK, NC 27709
Made in U.S.A. 542402
Alcohol 10%, Methylparaben 0.1% (added as a preservative)
For indications, dosage, precautions, etc., see accompanying package insert.
Store at 15° to 25°C (59° to 77°F) and protect from light.
CAUTION: Federal law prohibits dispensing without prescription.

a. What is the dose parameter according to the infant's weight?

____

____

b. Is the dose within safe dose parameters? ______________

Explain: ______________________________

c. How many milliliters would you give the child per dose?

____

126. Child has a urinary tract infection.
Order: Sulfisoxazole (Gastrisin) 300 mg, PO, q6h
Child's weight: 33 lbs
Pediatric dose parameter: 120–150 mg/kg/d; maximum: 6 g/d
Drug available: Gantrisin 500 mg/5 mL

a. How many kilograms does the child weigh? ____________________

b. What is the dose parameter according to the child's weight? ____________________

____

c. Is the dose within safe dose parameters? ____________________

Explain: ____________________

d. How many milliliters would you give the child per dose? ____________________

127. Child has urinary tract infection.
Order: Sulfisoxazole (Gastrisin) 0.75 g, PO, q6h
Child's height and weight: 56 inches, 75 lbs
Pediatric dose parameter: 4 g/m^2 in four divided doses
Drug available: Gantrisin 500 mg/5 mL

a. What is the child's body surface area (BSA, m^2)? (Use nomogram.) ____________________

b. What is the dose parameter according to the child's BSA? ____________________

____

c. Is the dose within safe dose parameters? ____________________

d. How many milliliters of Gantrisin would you give the child per dose? ____________________

____

128. Child has a urinary tract infection.
Order: Trimethoprim (TMP)/sulfamethoxazole (SMZ) (Septra) 60/300 mg, PO, q12h
Child's weight: 15 kg
Pediatric dose: TMP 8 mg/ SMZ 40 mg/kg/d in two divided doses
Drug available:

NDC 0081-0855-03
Suspension
(trimethoprim and sulfamethoxazole)
Each 5 mL (1 teaspoonful) contains trimethoprim 40 mg, sulfamethoxazole 200 mg, alcohol 0.26% and added as preservatives methylparaben 0.1%, sodium benzoate 0.1%.
CAUTION: Federal law prohibits dispensing without prescription.
CHERRY FLAVOR
Wellcome
BURROUGHS WELLCOME CO.
Research Triangle Park, NC 27709
Made in U.S.A. 595288
LOT
EXP
For indications, dosage, precautions, etc., see accompanying package insert.
Store at 15° to 25°C (59° to 77°F).
6505-01-136-8178

a. What is the dose parameter according to the child's weight?

____

____

b. Is the dose within safe dose parameters?

____

____

c. How many milliliters would you give the child per dose?

____

129. Child has a respiratory infection.
Order: Amoxicillin/clavulanate (Augmentin) 0.25 g, PO, q8h
Child's weight: 60 lbs
Pediatric dose parameter: 20–40 mg/kg/d in three divided doses
Drug available:

AUGMENTIN®

Tear along perforation
NSN 6505-01-207-8205
Directions for mixing:
Tap bottle until all powder flows freely. Add approximately 2/3 of total water for reconstitution **(total = 65 mL);** shake vigorously to wet powder. Add remaining water; again shake vigorously.
Dosage: See accompanying prescribing information.
Tear along perforation

Keep tightly closed.
Shake well before using.
Must be refrigerated.
Discard after 10 days.

AUGMENTIN®
AMOXICILLIN/
CLAVULANATE POTASSIUM
FOR ORAL SUSPENSION
When reconstituted, each 5 mL contains:
AMOXICILLIN, 250 MG,
as the trihydrate
CLAVULANIC ACID, 62.5 MG,
as clavulanate potassium
75mL *(when reconstituted)*

SB SmithKline Beecham

Use only if inner seal is intact.
Net contents: Equivalent to 3.75 g amoxicillin and 0.938 g clavulanic acid.
Store dry powder at room temperature.
Caution: Federal law prohibits dispensing without prescription.
SmithKline Beecham Pharmaceuticals
Philadelphia, PA 19101

EXP.

LOT

9405844-C

3 0029-6090-39 7

a. How many kilograms does the child weigh? ______

b. Is the dose within safe dose parameters? ______

Explain: ______

c. How many milliliters of Augmentin would you give the child per dose? ______

130. Child has cancer.
Order: Chlorambucil (Leukeran) 4.5 mg/m^2 per single dose
Child's height and weight: 58 in., 92 lbs
Pediatric dose parameter: See order
Drug available: Leukeran 2 mg tablets

a. What is the child's body surface area (BSA, m^2)? (Use nomogram.) ______

b. Is the dose within safe dose parameters? ______

Explain: ______

c. How many tablet(s) would you give the child? ______

131. Child has pruritus.
Order: Diphenydramine (Benadryl) 25 mg, PO, q6h
Child's weight: 22 lbs
Pediatric dose parameter: 5 mg/kg/d
Drug available: Benadryl 12.5 mg/5 mL

a. How many kilograms does the child weigh ? ______

b. Is the dose within safe dose parameters? ______

Explain: ______

c. How many milliliters would you give per dose? ______

______

132. Child (3 year old) has fever:
Order: Acetaminophen (Tylenol) 120 mg, PO, q4h × 4 doses/day, prn for temperature >102°F
Child's weight: 15 kg
Pediatric dose: 120–160 mg t.i.d., q.i.d.; maximum: 480 mg/d
Drug available:

ORIGINAL
CHILDREN'S
TYLENOL®
acetaminophen
ELIXIR
Fast, effective relief of children's fever and pain
CHERRY FLAVOR
Alcohol Free
Aspirin Free
Ibuprofen Free
4 fl oz (120mL)
80 mg per 1/2 teaspoon
(160mg Per 5mL)

a. Is the dose within safe dose parameters? ____________

Explain: ____________

____

____

b. How many milliliters of Tylenol would you give per dose?

____

____

____

____

133. Child has *Varicella zoster* infection.
Order: Acyclovir (Zovirax) 400 mg, PO, q8h, × 7 days
Child's height and weight: 46 in., 55 lbs
Pediatric dose parameter: 500 mg/m^2, q8h
Drug available:

1 pint (473 mL)
NDC 0081-0953-96
ZOVIRAX®
(ACYCLOVIR)
Suspension
Each 5 mL (1 teaspoonful) contains acyclovir 200 mg and added as preservatives methylparaben 0.1% and propylparaben 0.02%.
SHAKE WELL BEFORE USING.
For indications, dosage, precautions, etc., see accompanying package insert.
Store at 15° to 25°C (59° to 77°F).
Dispense in tight container as defined in the U.S.P.
CAUTION: Federal law prohibits dispensing without prescription.
647404
Made in U.S.A. U.S. Patent No. 4199574
LOT
EXP

a. What is the child's body surface area (BSA, m^2)?

____

b. What is the dose parameter according to the child's BSA?

____

____

c. Is the dose within safe dose parameters? ____________

____

____

d. How many milliliters would you give the child per dose?

____

____

____

134. Child has seizures.
Order: Ethosuximide (Zarontin) 125 mg, PO, b.i.d.
Child's weight: 24 kg
Pediatric dose parameter: 20 mg/kg/d; maximum: 1.5 g/d
Drug available: Zarontin 250 mg/5 mL

a. Is the dose within safe dose parameters? ______

Explain: ______

b. How many milliliters of Zarontin would you give per dose? ______

135. Child has attention deficit disorder (ADD).
Order: Methylphenidate HCl (Ritalin) 7.5 mg, PO, q.d. (a.m.)
Drug available: Ritalin 5 mg and 10 mg tablets

a. Which bottle of Ritalin would you use? ______

b. How many tablet(s) would you give? ______

136. Child has attention deficit disorder (ADD).
Order: Methylphenidate HCl (Ritalin) 15 mg, PO, q.d.
Drug available: Ritalin 5 mg and 10 mg tablets.

a. Which bottle of Ritalin would you use? ______

b. How many tablet(s) would you give? ______

______

137. Child has systemic infection.
Order: Tobramycin (Nebcin) 15 mg, IV, q6h
Child's weight: 12 kg
Pediatric dose parameter: 6–7.5 mg/kg/d in three or four divided doses
Drug available:

NDC 0002-0501-01
2 mL VIAL No. 782
℞ Lilly
NEBCIN®
PEDIATRIC
TOBRAMYCIN
SULFATE
INJECTION
USP
Equiv. to Tobramycin
20 mg per 2 mL
Multiple Dose
For I.M. or I.V. Use
Must dilute for I.V. use.
YE 1170 AMX
Eli Lilly & Company
Indpls., IN 46285, U.S.A.
Exp. Date/Control No.

a. How many milligrams of Nebcin would the child receive per day? ______

b. What is the dose parameter according to the child's weight? ______

______

c. Is the dose within safe dose parameters? ______

d. How many milliliters (IV) of Nebcin would be mixed into the IV fluids? ______

______

138. Child has severe respiratory infection.
 Order: Kanamycin (Kantrex) 50 mg, IV, q8h
 Child's weight: 44 lbs
 Pediatric dose parameter: 15 mg/kg/d in two or three divided doses
 Drug available:

a. How many kilograms does the child weigh? ______

b. What is the dose parameter according to the child's weight? ______

______

c. Is the dose within safe dose parameters? ______

d. How many milliliters (IV) of drug would be mixed into the IV fluids? ______

______

______

______

139. Child has congestive heart failure.
 Order: Digoxin (Lanoxin) 0.25 mg, IV, loading dose (LD)
 Child's weight: 35 lbs, or 16 kg
 Pediatric dose parameter: 0.015–0.035 mg/kg over 5 min, LD
 Drug available:

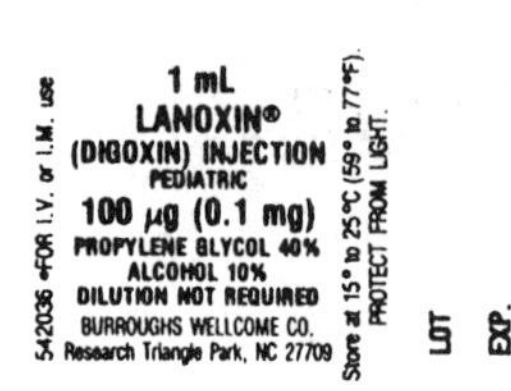

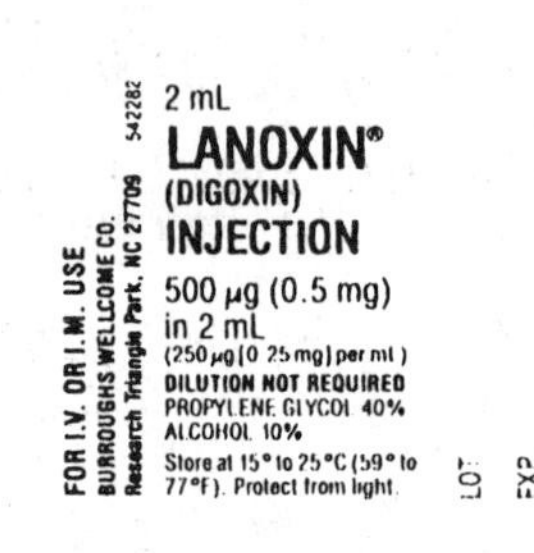

a. Which bottle of Lanoxin would you use? ______

b. What is the dose parameter according to the child's weight? ______

______

c. Is the dose within the safe dose parameters? ______

d. How many milliliters of Lanoxin would you give the child? ______

140. Child has a systemic infection.
Order: Oxacillin 200 mg, IV, q6h
Child's weight: 13 kg
Pediatric dose parameter: 50–100 mg/kg/d in four to six divided doses
Drug available:

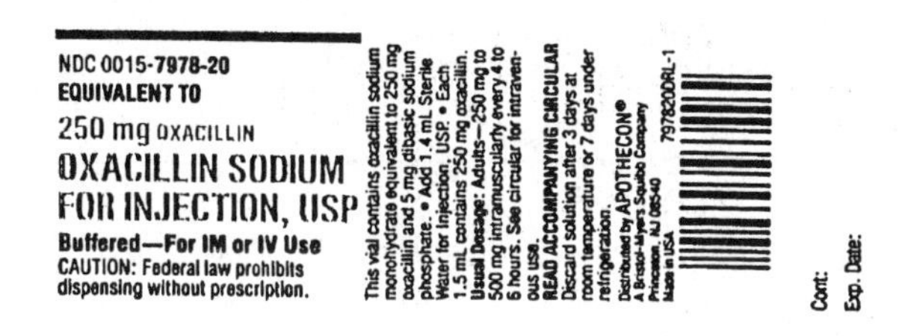

a. What is the dose parameter according to the child's weight? ______________________________

____

b. Is the dose within safe dose parameters? ______________________________

c. How many milliliters of diluent would you mix with Oxacillin? (see drug label)

____

____

d. How many milliliters (IV) of Oxacillin would be mixed into the IV fluids? ______________

____

____

141. Child has narcotic-induced respiratory depression.
Order: Naloxone HCl (Narcan) 0.32 mg, IM, stat
Child's weight: 32 kg
Pediatric dose parameter: 0.01 mg/kg, IM/IV, q2-3 min, prn
Drug available:

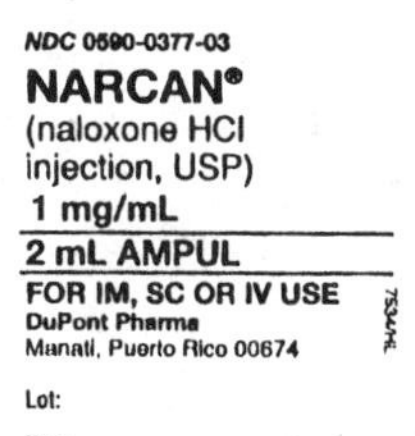

a. How many milliliters of Narcan would you give the child? ______________________________

DRUG CALCULATIONS BY BODY SURFACE AREA (BSA)

See adult and pediatric nomograms at the end of Part Three.

142. Give Dacarbazine 250 mg/m^2/day × 5 days
Patient height: 5'10"
Patient weight: 173 lbs
What is the daily dose?__________________________

143. Give 5-Flurouracil 450 mg/m^2/week
Patient height: 5'6"
Patient weight: 210 lbs
What is the weekly dose? __________________________

144. Give Leucovarin 200 mg/m^2/week
Patient height: 5'6"
Patient weight: 210 lbs
What is the weekly dose? __________________________

145. Give Cisplatin 30 mg/m^2/day × 3 days
Patient height: 72 in.
Patient weight: 80 kg
What is the daily dose?__________________________

146. Give Cisplatinum 80 mg/m^2/week
Patient height: 6'2"
Patient weight: 186 lbs
What is the weekly dose? __________________________

147. Give Etaposide 120 mg/m^2/week
Patient height: 74 in.
Patient weight: 70 kg
What is the weekly dose? __________________________

148. Give Cytoxan 600 mg/m^2/week
Patient height: 70 in.
Patient weight: 85 kg
What is the weekly dose? __________________________

149. Give Adriamycin 60 mg/m^2/week
Patient height: 70 in.
Patient weight: 85 kg
What is the weekly dose? __________________________

150. Give Vincristine 2 mg/m^2/week
Patient height: 62 in.
Patient weight: 75 kg

What is the weekly dose? ____________________

151. Give Mitomycin 15 mg/m^2/week
Patient height: 65 in.
Patient weight: 64 kg

What is the weekly dose? ____________________

152. Give Mitoxantrone 12 mg/m^2/day × 3 days
Patient height: 5'8"
Patient weight: 150 lbs

What is the daily dose?____________________

153. Give Cystosine arabinoside 100 mg/m^2/day × 7 days
Patient height: 5'2"
Patient weight: 130 lbs

What is the daily dose?____________________

154. Give Methotrexate 3.3mg/m^2/day × 7 days
Patient height: 72 in.
Patient weight: 82 kg

What is the daily dose?____________________

155. Give Prednisone 60 mg/m^2/day × 7 days
Patient height: 72 in.
Patient weight: 82 kg

What is the daily dose?____________________

156. Give Idarubicin hydrochloride 12 mg/m^2/day × 3 days
Patient height: 60 in.
Patient weight: 60 kg

What is the daily dose?____________________

157. Give Cytarabine 100 mg/m^2/day × 7 days
Patient height: 64 in.
Patient weight: 60 kg

What is the daily dose?____________________

INTRAVENOUS FLUIDS

Hours to Administer

How many hours are required to administer the following fluids at the given rates?

158. Rate: TPN 1000 mL at 30 mL/hr

 Hours? ______

159. Rate: D_5 1/4 1000 mL at 100 mL/hr

 Hours? ______

160. Rate: D_5 1/2 1000 mL at 125 mL/hr

 Hours? ______

161. Rate: D_5W 500 mL at 20 mL/hr

 Hours? ______

162. Rate: LRS 1000 mL at 90 mL/hr

 Hours? ______

163. Rate: $D_{10}W$ 1000 mL at 40 mL/hr

 Hours? ______

164. Rate: D_5W 1000 mL at 50 mL/hr

 Hours? ______

165. Rate: D_5LRS 1000 mL at 150 mL/hr

 Hours? ______

166. Rate: D_5 1/3 1000 mL at 60 mL/hr

 Hours? ______

167. Rate: NSS 1000 mL at 80 mL/hr

 Hours? ______

Direct IV Injection (IV Push)

For problems 168–187, calculate the number of milliliters needed for each intravenous dose for direct IV push. Use the basic formula method or the ratio and proportion method.

168. Atropine sulfate 0.6 mg, IV push
Available: 1 mg/10 mL

Amount (mL)?____________________

169. Lidocaine 80 mg, IV push
Available: 100 mg/10 mL

Amount (mL)?____________________

170. Lidocaine 75 mg, IV push
Available: 100 mg/10 mL

Amount (mL)?____________________

171. Epinephrine 0.75 mg, IV push
Available: 1:10,000

Amount (mL)?____________________

172. Atropine sulfate 0.5 mg, IV push
Available: 1 mg/10 mL

Amount (mL)?____________________

173. Midazolam 2 mg, IV push
Available: 5 mg/mL

Amount (mL)?____________________

174. Lasix 80 mg, IV push
Available: 100 mg/10 mL

Amount (mL)?____________________

175. Adenosine 6 mg, rapid IV push
Available: 3 mg/mL

Amount (mL)?____________________

176. Adenosine 12 mg, rapid IV push
Available: 3 mg/mL

Amount (mL)?____________________

177. Magnesium sulfate 3 g, IV push
Available: 1 g/mL

Amount (mL)?____________________

178. Gentamycin 75 mg, IV push
Available: 40 mg/mL
Amount (mL)?______________________________

179. Methylprednisone sodium 80 mg, IV push
Available: 125 mg/mL
Amount (mL)?______________________________

180. Verapamil HCl 0.15 mg/kg, IV push
Patient weight: 75 kg
Available: 2.5 mg/mL
Amount (mL)?______________________________

181. Metoprolol tartrate 5 mg, IV push
Available: 1 mg/mL
Amount (mL)?______________________________

182. Hydralazine HCl 15 mg, slow IV push
Available: 20 mg/mL
Amount (mL)?______________________________

183. Flumazenil 0.2 mg, IV push
Available: 0.1 mg/mL
Amount (mL)?______________________________

184. Sodium bicarbonate 1 mEq/kg, IV push
Patient weight: 70 kg
Available: 44.6 mEq/50 mL
Amount (mL)?______________________________

185. Digoxin 1 mg, IV push
Available: 0.5 mg/2 mL
Amount (mL)?______________________________

186. Digibind 800 mg, IV push
Available: 10 mg/mL after reconstitution
Amount (mL)?______________________________

187. Digoxin 0.125 mg, IV push
Available: 0.5 mg/2 mL
Amount (mL)?______________________________

For problems 188–190, answer the following questions:

a. How many milliliters are needed for each IV dose (push)?
b. The IV drug dose should be administered over how many minutes?

188. Phenytoin 900 mg, IV push at 50 mg/min
Available: 50 mg/mL

a. Amount (mL)?______________________________

b. Minutes? ______________________________

189. Phenobarbital 10 mg/kg, IV push at 50 mg/min
Patient weight: 65 kg
Available: 30 mg/mL

a. Amount (mL)?______________________________

b. Minutes? ______________________________

190. Amrinone 0.75 mg/kg, IV push over 3 min
Patient weight: 75 kg
Available: 5 mg/mL

a. Amount (mL)?______________________________

b. Minutes? ______________________________

IV Drop Rate

What are the drop rates per minute for the following drop factors?

191. $D_{10}W$ at 40 mL/hr

a. 10 gtt/mL?______________________________

b. 15 gtt/mL?______________________________

c. 20 gtt/mL?______________________________

d. 60 gtt/mL?______________________________

192. D_5LRS at 100 mL/hr

a. 10 gtt/mL?______________________________

b. 15 gtt/mL?______________________________

c. 20 gtt/mL?______________________________

d. 60 gtt/mL?______________________________

193. D_5W at 80 mL/hr

a. 10 gtt/mL?______________________________

b. 15 gtt/mL?______________________________

c. 20 gtt/mL?______________________________

d. 60 gtt/mL?______________________________

194. D_5 1/4 NS at 125 mL/hr
   a. 10 gtt/mL?______
   b. 15 gtt/mL?______
   c. 20 gtt/mL?______
   d. 60 gtt/mL?______

195. D_5LRS at 150 mL/hr
   a. 10 gtt/mL?______
   b. 15 gtt/mL?______
   c. 20 gtt/mL?______
   d. 60 gtt/mL?______

196. NSS at 200 mL/hr
   a. 10 gttt/mL? ______
   b. 15 gtt/mL?______
   c. 20 gtt/mL?______
   d. 60 gtt/mL?______

197. D_5 1/2 at 75 mL/hr
   a. 10 gtt/mL?______
   b. 15 gtt/mL?______
   c. 20 gtt/mL?______
   d. 60 gtt/mL?______

198. PSS at 500 mL/hr
   a. 10 gtt/mL?______
   b. 15 gtt/mL?______
   c. 20 gtt/mL?______
   d. 60 gtt/mL?______

199. D_5W at 30 mL/hr
   a. 10 gtt/mL?______
   b. 15 gtt/mL?______
   c. 20 gtt/mL?______
   d. 60 gtt/mL?______

200. D_5W at 60 mL/hr
   a. 10 gtt/mL?__________
   b. 15 gtt/mL?__________
   c. 20 gtt/mL?__________
   d. 60 gtt/mL?__________

201. D_5 1/4 NS at 150 mL/hr
   a. 10 gtt/mL?__________
   b. 15 gtt/mL?__________
   c. 20 gtt/mL?__________
   d. 60 gtt/mL?__________

202. $D_{10}W$ at 50 mL/hr
   a. 10 gtt/mL?__________
   b. 15 gtt/mL?__________
   c. 20 gtt/mL?__________
   d. 60 gtt/mL?__________

203. NSS at 1000 mL/hr
   a. 10 gtt/mL?__________
   b. 15 gtt/mL?__________
   c. 20 gtt/mL?__________
   d. 60 gtt/mL?__________

204. D_5W at 20 mL/hr
   a. 10 gtt/mL?__________
   b. 15 gtt/mL?__________
   c. 20 gtt/mL?__________
   d. 60 gtt/mL?__________

205. LRS at 90 mL/hr
   a. 10 gtt/mL?__________
   b. 15 gtt/mL?__________
   c. 20 gtt/mL?__________
   d. 60 gtt/mL?__________

Antibiotics

For problems 206–237, answer these questions:

a. What is the correct drug dose?
b. What is the flow rate in gtt/min?

206. Give Amikacin 800 mg IV, now
Available: Amikacin 500 mg/2 mL
Set and solution: D_5W 150 mL bag and a drop factor of 15 gtt/mL set
Instructions: Infuse over 60 min

a. Drug dose?____________________

b. Flow rate? ____________________

207. Give Amphotericin B 20 mg IV daily
Available: Amphotericin B 50 mg/15 mL
Set and solution: D_5W 500 mL bag and a drop factor of 10gtt/mL set
Instructions: Infuse over 6 hr

a. Drug dose?____________________

b. Flow rate? ____________________

208. Give Ampicillin 500 mg IV q6h
Available: Ampicillin 1 g, add 4.5 mL of diluent to yield 5mL
Set and solution: Buretrol with a drop factor of 60 gtt/mL and NSS 500 mL
Instructions: Dilute drug in 50 mL and infuse over 15 min

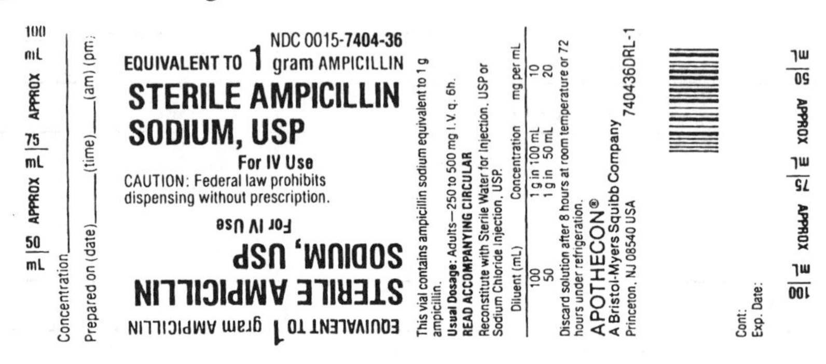

a. Drug dose?____________________

b. Flow rate? ____________________

209. Give Cefazolin 1g IV q8h
Available: Cefazolin 1 g, add 4.5 mL of diluent to yield 5 mL
Set and solution: Buretrol with a drop factor of 60 gtt/mL and D_5W 500 mL
Instructions: Dilute drug in 50 mL and infuse over 40 min

a. Drug dose?____________________

b. Flow rate? ____________________

210. Give Aztreonam 1 g IV q8h
Available: Aztreonam 1 g, add 4.5 mL of diluent to yield 5 mL
Set and solution: Buretrol with a drop factor of 60 gtt/mL and D_5W 500 mL
Instruction: Dilute drug in 50 mL and infuse over 20 min

a. Drug dose?____________________

b. Flow rate? ____________________

211. Give Bactrim 400 mg IV q6h
Available: Bactrim 400 mg /5 mL
Set and solution: Buretrol with a drop factor of 60 gtt/mL and D_5W 500 mL
Instructions: Dilute drug in 125 mL and infuse over 90 min

a. Drug dose?____________________

b. Flow rate? ____________________

212. Give Cefotetan 2g IV q8h
Available: Cefotetan 1 g, add 4.5 mL of diluent to yield 5 mL
Set and solution: Buretrol with a drop factor of 60 gtt/mL and D5W 500 mL.
Instructions: Dilute drug in 75 mL and infuse over 30 min

a. Drug dose?____________________

b. Flow rate? ____________________

213. Give Clindamycin 300 mg IV q8h
Available: Clindamycin 600 mg /4 mL
Set and solution: 100 mL bag of D_5W and a 20 gtt/mL set
Instructions: Infuse over 20 min

Sterile NDC 0205-2801-18
Clindamycin Phosphate
Injection, USP
Equivalent to 600 mg/4 mL clindamycin (150 mg/mL).
FOR IM OR IV* USE.
*If given intravenously, must be diluted before use.
4 mL Single Dose Vial
Control Exp.
SPECIMEN
LEDERLE PARENTERALS, INC.
Carolina, Puerto Rico 00987

a. Drug dose?____________________

b. Flow rate? ____________________

214. Give Clindamycin 900 mg IV q12h
Available: Clindamycin 600 mg/4 mL
Set and solution: 100 mL bag of D_5W and a 10 gtt/mL set
Instructions: Infuse over 30 min

a. Drug dose?____________________

b. Flow rate? ____________________

215. Give Erthromycin 500 mg IV q6h
Available: Erthromycin 1g, add 4.5 mL diluent to yield 5 mL
Set and solution: Buretrol with a 60 gtt/ml drop factor and NSS 500 mL
Instructions: Dilute in 100 mL and infuse over 60 min

a. Drug dose?____________________

b. Flow rate? ____________________

216. Give Ceftazidime 2g q8h
Available: Ceftazidime 1g, add 4.5 mL diluent to yield 5 mL
Set and solution: 100 mL bag of NSS and a set with a drop factor of 20 gtt/mL
Instructions: Dilute in 100 mL NSS and infuse over 30 min

a. Drug dose?____________________

b. Flow rate? ____________________

217. Give Gentamycin 60 mg q12h
Available: Gentamycin 80 mg/2 mL
Set and solution: Buretrol with a 60 gtt/mL drop factor and D_5W 500 mL
Instructions: Dilute in 50 mL and infuse over 30 min

a. Drug dose?____________________

b. Flow rate? ____________________

218. Give Gentamycin 120 mg q12h
Available: Gentamycin 80 mg/2 mL
Set and solution: 200 mL bag NSS and a 20gtt/mL drop factor set
Instructions: Infuse over 60 min

a. Drug dose?____________________

b. Flow rate? ____________________

219. Give Nafcillin 250 mg q4h
Available: Nafcillin 500 mg and 1.8 mL diluent to yield 2 mL
Set and solution: Buretrol with a 60 gtt/mL drop factor and NSS 500 mL
Instructions: Dilute in 50 mL and infuse over 30 min

NDC 0015-7224-20
EQUIVALENT TO
500 mg NAFCILLIN
NAFCILLIN SODIUM FOR INJECTION, USP
Buffered—For IM or IV Use
CAUTION: Federal law prohibits dispensing without prescription.

When reconstituted with 1.8 mL diluent, (SEE INSERT-INTRAMUSCULAR ROUTE), each vial contains 2 mL solution. Each mL of solution contains nafcillin sodium, as the monohydrate, equivalent to 250 mg nafcillin, buffered with 10 mg sodium citrate. Read accompanying circular for complete stability data. Usual Dosage: Adults—500 mg every 4 to 6 hours. Read accompanying circular for directions for IM or IV use.
APOTHECON®
A Bristol-Myers Squibb Company
Princeton, NJ 08540 USA
722420DRL-2

Cont:
Exp. Date:

a. Drug dose?____________________

b. Flow rate? ____________________

220. Give Nafcillin 1 g q4h
Available: Nafcillin 1 g, add 2 mL diluent to yield 3 mL
Set and solution: D_5W 50 mL bag and a set with a 60 gtt/mL drop factor
Instructions: Infuse over 60 min

NDC 0015-7195-28
EQUIVALENT TO 1 gram NAFCILLIN
NAFCILLIN SODIUM
FOR INJECTION, USP
Buffered—For IV Use
CAUTION: Federal law prohibits dispensing without prescription.
Concentration ______ Prepared on (date) ______ (time) ______ (am) (pm)
This vial contains nafcillin sodium, as the monohydrate, equivalent to 1 g nafcillin buffered with 40 mg sodium citrate.
Usual Dosage: Adults—500 mg every 4 to 6 hours.
READ ACCOMPANYING CIRCULAR
Reconstitute with Sterile Water for Injection, USP or 0.9% Sodium Chloride Injection, USP. The table gives concentrations for various diluent volumes.

Diluent Added (mL)	Solution Concentration	mg per mL
100	1 g in 100 mL	10
50	1 g in 50 mL	20
20	1 g in 20 mL	50

Read accompanying circular for complete stability data.
APOTHECON®
A Bristol-Myers Squibb Company
Princeton, NJ 08540 USA
719528DRL-1
Cont:
Exp. Date:

a. Drug dose? ______________________

b. Flow rate? ______________________

221. Give Penicillin G Aqueous 5,000,000 units q4h
Available: Penicillin G Aqueous 10,000,000 units, add 10 mL diluent to yield 12 mL
Set and solution: D_5W 100 mL bag and a set with a 20 gtt/mL drop factor
Instructions: Infuse over 40 min

a. Drug dose? ______________________

b. Flow rate? ______________________

222. Give Imipenem-Cilastatin 500 mg q6h
Available: Imipenem-Cilastatin 1g, add 2 mL diluent to yield 3 mL
Set and solution: D_5W 100 mL bag and a 10 gtt/mL drop factor
Instructions: Infuse over 20 min

a. Drug dose? ______________________

b. Flow rate? ______________________

223. Give Ceftriaxone 2 g q12h
Available: Ceftriaxone 1 g, add 5 mL to each vial
Set and solution: Buretrol with a 60 gtt/mL drop factor and NSS 500 mL bag
Instructions: Dilute with 75 mL and infuse over 30 min

a. Drug dose? ______________________

b. Flow rate? ______________________

224. Give Tetracycline 250 mg IV q6h
Available: Tetracycline 500 mg, add 1.8 mL diluent to yield 2 mL
Set and solution: 100 mL bag of D_5W and a set with a 20 gtt/mL drop factor
Instructions: Infuse over 60 min

a. Drug dose? ______________________

b. Flow rate? ______________________

225. Give Tobramycin 60 mg IV q6h
Available: Tobramycin 40 mg/2 mL
Set and solution: 50 mL bag D_5Wand a set with a 15 gtt/mL drop factor
Instructions: Dilute with 50 mL and infuse over 30 min
a. Drug dose?____________________
b. Flow rate? ____________________

226. Give Ticarcillin 4 g IV q6h
Available: Ticarcillin 6g, add 8 mL diluent to yield 10 mL
Set and solution: 50 mL bag of D_5W and a set with a 10 gtt/mL drop factor
Instructions: Infuse over 30 min
a. Drug dose?____________________
b. Flow rate? ____________________

227. Give Vancomycin 500 mg q6h
Available: Vancomycin 500 mg, add 4.5 mL diluent to yield 5 mL
Set and solution: 100 mL bag of D_5W and a set with a 10 gtt/mL drop factor
Instructions: Infuse over 60 min
a. Drug dose?____________________
b. Flow rate? ____________________

228. Give Vancomycin 1.5 g q12h
Available: Vancomycin 500 mg, add 4.5 mL to each vial to yield 5 mL/vial
Set and solution: 250 mL bag of D_5W and a set with a 10 gtt/mL drop factor
Instructions: Infuse over 2 hr
a. Drug dose?____________________
b. Flow rate? ____________________

229. Give Rifampin 600 mg daily
Available: Rifampin 600 mg, add 5 mL diluent
Set and solution: 500 mL bag of D_5W and a set with a 15 gtt/mL drop factor
Instructions: Infuse over 3 hr
a. Drug dose?____________________
b. Flow rate? ____________________

230. Give Cefotaxime 2g q6h
Available: Cefotaxime 1g, add 5 mL diluent to each vial
Set and solution: 100 mL bag of D_5W and a set with a 15 gtt/mL drop factor
Instructions: Infuse over 30 min
a. Drug dose?____________________
b. Flow rate? ____________________

231. Give Cephalothin 2g q6h
Available: Cephalothin 1g, add 5 mL diluent to each vial
Set and solution: Buretrol with a 60 gtt/mL drop factor and D_5W 500 mL
Instructions: Dilute with 75 mL and infuse over 30 min

a. Drug dose? ______

b. Flow rate? ______

232. Give Cefuroxime 750 mg q8h
Available: Cefuroxime 1g, add 4.5 mL diluent to yield 5 mL
Set and solution: Buretrol with a 60 gtt/mL drop factor and D_5W 500 mL
Instructions: Dilute with 50 mL and give over 30 min

a. Drug dose? ______

b. Flow rate? ______

233. Give Cephapirin 500 mg q4h
Available: Cephapirin 2 g, add 10 mL diluent to yield 5 mL/g
Set and solution: 100 mL bag of D_5W and a set with a 20 gtt/mL drop factor
Instructions: Infuse over 20 min

Concentration
Prepared on (Date) (A.M.) (P.M.)
(Time)
NDC 0015-7629-30
EQUIVALENT TO
2 gram CEPHAPIRIN
CEFADYL®
Sterile Cephapirin Sodium, USP
For IV Use
CAUTION: Federal law prohibits dispensing without prescription
APOTHECON®
This vial contains cephapirin sodium equivalent to 2 g cephapirin. Nitrogen in headspace.
Usual Dosage: Adults—4 to 12 g daily, intravenously. For IV use, add 10 mL Sterile Water for Injection. Each 5 mL will contain 1 g cephapirin.
Following reconstitution, the solution is stable for 12 hours at room temperature.
READ ACCOMPANYING CIRCULAR
Distributed by APOTHECON®
A Bristol-Myers Squibb Company
Princeton, NJ 08540
Made in USA
762930DRL-1

a. Drug dose? ______

b. Flow rate? ______

234. Give Cephradine 1g q6h
Available: Cephradine 500 mg, add 4.5 mL diluent to each vial to yield 5 mL/vial
Set and solution: 50 mL bag of D_5W and a set with a 15 gtt/mL drop factor
Instructions: Infuse over 30 min

Intramuscular or intravenous
Vial contains 500 mg cephradine with 157 mg anhydrous sodium carbonate: total sodium content is approx. 68 mg (3 mEq)
Protect from light
Usual adult dosage: 500 mg to 1 gram qid—See insert
SQUIBB®
NDC 0003-1198-10
500 mg
VELOSEF®
Cephradine for Injection USP
Caution: Federal law prohibits dispensing without prescription
Use solution within 2 hr. if stored at room temperature or within 24 hr. if stored at 5° C.
Manufactured by Von Heyden G.m.b.H. Regensburg, Germany for E.R. Squibb & Sons, Inc. Princeton, NJ 08540 USA
C2737A / A9810
EXP. SEPT 1, 93
0J31821

a. Drug dose? ______

b. Flow rate? ______

235. Give Cefoperazone 2 g q12h
Available: Cefoperazone 1 g, add 4.5 mL to each vial to yield 5 mL/vial
Set and solution: 50 mL bag of D_5W and a set with a 15 gtt/mL drop factor
Instructions: Infuse over 30 min

a. Drug dose?____________________

b. Flow rate? ____________________

236. Give Oxacillin 250 mg IV q4h
Available: Oxacillin 1 g, add 5.7 mL to yield 6 mL
Set and solution: 100 mL bag of D_5W and a set with a 20 gtt/mL drop factor
Instructions: Infuse over 30 min

NDC 0015-7981-20
EQUIVALENT TO
1 gram OXACILLIN
OXACILLIN SODIUM FOR INJECTION, USP
Buffered—For IM or IV Use
CAUTION: Federal law prohibits dispensing without prescription.
APOTHECON® A BRISTOL-MYERS SQUIBB COMPANY

This vial contains oxacillin sodium monohydrate equivalent to 1 gram oxacillin and 20 mg dibasic sodium phosphate. Add 5.7 mL Sterile Water for Injection, USP. • Each 1.5 mL contains 250 mg oxacillin. • Usual Dosage: Adults—250 mg to 500 mg intramuscularly every 4 to 6 hours. See circular for intravenous use. READ ACCOMPANYING CIRCULAR. Discard solution after 3 days at room temperature or 7 days under refrigeration.
APOTHECON®
A Bristol-Myers Squibb Company
Princeton, NJ 08540 USA
798120DRL-2

Cont:
Exp. Date:

a. Drug dose?____________________

b. Flow rate? ____________________

237. Give Minocycline 200 mg IV, now
Available: Minocycline 100 mg, add 2 mL diluent per vial
Set and solution: 500 mL NSS and a set with a drop factor of 10 gtt/mL
Instructions: Infuse over 1 hr

6505-00-149-0574 NDC 0205-5305-94
Minocin®
Sterile Minocycline Hydrochloride
Intravenous 100 mg
CAUTION: Federal law prohibits dispensing without prescription.
Contains: Sterile Minocycline HCl equivalent to 100 mg Minocycline. Hydrochloric acid to pH approx. 2.4
Usual Adult Dose: 100 mg Intravenously every 12 hours. See Enclosed Circular.
100 mg Vial 23149 DX54
Control No.
Exp. Date
LEDERLE PARENTERALS, INC.
Carolina, Puerto Rico 00630

a. Drug dose?____________________

b. Flow rate? ____________________

Critical Care: Infusion Rates

For problems 238–279, answer these questions (questions cannot always be answered in sequence):

a. What is the concentration of the solution?
b. What is the volume per minute?
c. What is the volume per hour?
d. What is the concentration per minute?
e. What is the concentration per hour?

238. Aminophylline 1 g in 250 mL D_5W at 20 mg/hr

a. Conc of soln? __________
b. Vol/min? __________
c. Vol/hr? __________
d. Conc/min? __________
e. Conc/hr? __________

239. Lidocaine 2 g in 500 mL D_5W at 3 mg/min

a. Conc of soln? __________
b. Vol/min? __________
c. Vol/hr? __________
d. Conc/min? __________
e. Conc/hr? __________

240. Dopamine 400 mg in 250 mL D_5W at 3 mcg/kg/min; patient weight = 60 kg

a. Conc of soln? __________
b. Vol/min? __________
c. Vol/hr? __________
d. Conc/min? __________
e. Conc/hr? __________

241. Nitroglycerine 50 mg in 250 mL D_5W at 10 mcg/min

a. Conc of soln? __________
b. Vol/min? __________
c. Vol/hr? __________
d. Conc/min? __________
e. Conc/hr? __________

242. Nitroglycerine 100 mg in 250 mL D_5W at 25 mcg/min

a. Conc of soln? ______
b. Vol/min? ______
c. Vol/hr? ______
d. Conc/min? ______
e. Conc/hr? ______

243. Procainamide 1 g in 500 mL D_5W at 4 mg/min

a. Conc of soln? ______
b. Vol/min? ______
c. Vol/hr? ______
d. Conc/min? ______
e. Conc/hr? ______

244. Nitroprusside 50 mg in 1000 mL D_5W at 5 mcg/kg/min; patient weight = 60 kg

a. Conc of soln? ______
b. Vol/min? ______
c. Vol/hr? ______
d. Conc/min? ______
e. Conc/hr? ______

245. Isoproterenol 1 mg in 500 mL D_5W at 5 mcg/min

a. Conc of soln? ______
b. Vol/min? ______
c. Vol/hr? ______
d. Conc/min? ______
e. Conc/hr? ______

246. Isoproterenol 2 mg in 500 mL D_5W at 5 mcg/min

a. Conc of soln? ______
b. Vol/min? ______
c. Vol/hr? ______
d. Conc/min? ______
e. Conc/hr? ______

247. Norepinephrine 4 mg in 1000 mL D_5W at 10 mcg/min
   a. Conc of soln? ______
   b. Vol/min? ______
   c. Vol/hr? ______
   d. Conc/min? ______
   e. Conc/hr? ______

248. Norepinephrine 4 mg in 500 mL D_5W at 8 mcg/min
   a. Conc of soln? ______
   b. Vol/min? ______
   c. Vol/hr? ______
   d. Conc/min? ______
   e. Conc/hr? ______

249. Epinephrine 5 mg in 250 mL D_5W at 4 mcg/min
   a. Conc of soln? ______
   b. Vol/min? ______
   c. Vol/hr? ______
   d. Conc/min? ______
   e. Conc/hr? ______

250. Bretylium tosylate 1 g in 250 mL D_5W at 2 mg/min
   a. Conc of soln? ______
   b. Vol/min? ______
   c. Vol/hr? ______
   d. Conc/min? ______
   e. Conc/hr? ______

251. Phenylephrine HCl 10 mg in 500 mL D_5W at 150 mcg/min
   a. Conc of soln? ______
   b. Vol/min? ______
   c. Vol/hr? ______
   d. Conc/min? ______
   e. Conc/hr? ______

252. Regular humulin insulin 100 units in 1000 mL PSS at 6 units/hr
   a. Conc of soln? ______
   b. Vol/min? ______
   c. Vol/hr? ______
   d. Conc/min? ______
   e. Conc/hr? ______

253. Oxytocin 10 units in 1000 mL PSS at 10 mU/min
   a. Conc of soln? ______
   b. Vol/min? ______
   c. Vol/hr? ______
   d. Conc/min? ______
   e. Conc/hr? ______

254. Esmolol 2.5 g in 500 mL D_5W at 100 mcg/kg/min; patient weight = 100 kg
   a. Conc of soln? ______
   b. Vol/min? ______
   c. Vol/hr? ______
   d. Conc/min? ______
   e. Conc/hr? ______

255. Esmolol 2.5 g in 250 mL D_5W at 200 mcg/kg/min; patient weight = 50 kg
   a. Conc of soln? ______
   b. Vol/min? ______
   c. Vol/hr? ______
   d. Conc/min? ______
   e. Conc/hr? ______

256. Phenylephrine HCl 10 mg in 500 mL D_5W at 50 mcg/min
   a. Conc of soln? ______
   b. Vol/min? ______
   c. Vol/hr? ______
   d. Conc/min? ______
   e. Conc/hr? ______

257. Morphine sulfate 100 mg in 250 mL D_5W at 3 mg/hr
   a. Conc of soln? ______
   b. Vol/min? ______
   c. Vol/hr? ______
   d. Conc/min? ______
   e. Conc/hr? ______

258. Dobutamine 500 mg in 500 mL D_5W at 5 mcg/kg/min; patient weight = 80 kg
   a. Conc of soln? ______
   b. Vol/min? ______
   c. Vol/hr? ______
   d. Conc/min? ______
   e. Conc/hr? ______

259. Dobutamine 250 mg in 500 mL D_5W at 10 mcg/kg/min; patient weight = 100 kg
   a. Conc of soln? ______
   b. Vol/min? ______
   c. Vol/hr? ______
   d. Conc/min? ______
   e. Conc/hr? ______

260. Ranitidine HCl 150 mg in 250 mL D_5W at 6.25 mg/hr
   a. Conc of soln? ______
   b. Vol/min? ______
   c. Vol/hr? ______
   d. Conc/min? ______
   e. Conc/hr? ______

261. Heparin 12,500 units in 250 mL D_5W at 1200 units/hr
   a. Conc of soln? ______
   b. Vol/min? ______
   c. Vol/hr? ______
   d. Conc/min? ______
   e. Conc/hr? ______

262. Diltiazem HCl in 125 mg in 250 mL PSS at 5 mg/hr
   a. Conc of soln? ____________________
   b. Vol/min? ____________________
   c. Vol/hr? ____________________
   d. Conc/min? ____________________
   e. Conc/hr? ____________________

263. Aminophylline 1 g in 250 mL D_5W at 16 mg/hr
   a. Conc of soln? ____________________
   b. Vol/min? ____________________
   c. Vol/hr? ____________________
   d. Conc/min? ____________________
   e. Conc/hr? ____________________

264. Lidocaine HCl 2 g in 500 mL D_5W at 2 mg/min
   a. Conc of soln? ____________________
   b. Vol/min? ____________________
   c. Vol/hr? ____________________
   d. Conc/min? ____________________
   e. Conc/hr? ____________________

265. Lidocaine HCl 2 g in 500 mL D_5W at 3 mg/min
   a. Conc of soln? ____________________
   b. Vol/min? ____________________
   c. Vol/hr? ____________________
   d. Conc/min? ____________________
   e. Conc/hr? ____________________

266. Dopamine 400 mg in 250 mL D_5W at 5 mcg/kg/min; patient weight = 165 lbs
   a. Conc of soln? ____________________
   b. Vol/min? ____________________
   c. Vol/hr? ____________________
   d. Conc/min? ____________________
   e. Conc/hr? ____________________

267. Heparin 25,000 units in 250 mL D_5W at 1000 units/hr

a. Conc of soln? ____________________

b. Vol/min? ____________________

c. Vol/hr? ____________________

d. Conc/min? ____________________

e. Conc/hr? ____________________

268. Nitroglycerine 50 mg in 250 mL D_5W at 25 mcg/min

a. Conc of soln? ____________________

b. Vol/min? ____________________

c. Vol/hr? ____________________

d. Conc/min? ____________________

e. Conc/hr? ____________________

269. Vasopressin 100 units in 250 mL D_5W at 0.2 units/min

a. Conc of soln? ____________________

b. Vol/min? ____________________

c. Vol/hr? ____________________

d. Conc/min? ____________________

e. Conc/hr? ____________________

270. Procainamide HCl 2 g in 250 mL D_5W at 3 mg/min

a. Conc of soln? ____________________

b. Vol/min? ____________________

c. Vol/hr? ____________________

d. Conc/min? ____________________

e. Conc/hr? ____________________

271. Nitroprusside 50 mg in 500 mL D_5W at 3 mcg/kg/min; patient weight = 70 kg

a. Conc of soln? ____________________

b. Vol/min? ____________________

c. Vol/hr? ____________________

d. Conc/min? ____________________

e. Conc/hr? ____________________

272. Procainamide HCl 2 g in 250 mL D_5W at 15 mg/min

a. Conc of soln? ______

b. Vol/min? ______

c. Vol/hr? ______

d. Conc/min? ______

e. Conc/hr? ______

273. Heparin 25,000 units in 250 mL D_5W at 1400 units/hr

a. Conc of soln? ______

b. Vol/min? ______

c. Vol/hr? ______

d. Conc/min? ______

e. Conc/hr? ______

274. Atracrium besylate 20 mg in 100 mL D_5W at 5 mcg/kg/min; patient weight = 80 kg

a. Conc of soln? ______

b. Vol/min? ______

c. Vol/hr? ______

d. Conc/min? ______

e. Conc/hr? ______

275. Norepinephrine 4 mg in 250 mL D_5W at 2 mcg/min

a. Conc of soln? ______

b. Vol/min? ______

c. Vol/hr? ______

d. Conc/min? ______

e. Conc/hr? ______

276. Diltiazem 250 mg in 250 mL D_5W at 5 mg/hr

a. Conc of soln? ______

b. Vol/min? ______

c. Vol/hr? ______

d. Conc/min? ______

e. Conc/hr? ______

277. Morphine sulfate 100 mg in 500 mL D_5W at 8 mg/hr
   a. Conc of soln? ______
   b. Vol/min? ______
   c. Vol/hr? ______
   d. Conc/min? ______
   e. Conc/hr? ______

278. Alteplase 100 mg in 250 mL D_5W at 20 mg/hr
   a. Conc of soln? ______
   b. Vol/min? ______
   c. Vol/hr? ______
   d. Conc/min? ______
   e. Conc/hr? ______

279. Potassium chloride 80 mEq in 100 mL D_5W at 10 mEq/hr
   a. Conc of soln? ______
   b. Vol/min? ______
   c. Vol/hr? ______
   d. Conc/min? ______
   e. Conc/hr? ______

Ob/Gyn IV Drugs

For problems 280–287, answer these questions:

a. What is the concentration of the solution?
b. What is the concentration per hour?
c. What is the volume per hour?

280. Order: Pitocin 10 units in NSS 500 mL, infuse at 10 mU/min
   Available: Oxytocin 10 units/mL
   a. Conc of soln? ______
   b. Conc/hr? ______
   c. Vol/hr? ______

281. Order: Pitocin 20 units in NSS 500 mL, infuse at 20 mU/min
   Available: Oxytocin 10 units/mL
   a. Conc of soln? ______
   b. Conc/hr? ______
   c. Vol/hr? ______

282. Order: Pitocin 40 units in NSS 1000 mL, infuse at 10 mU/min
Available: Oxytocin 10 units/mL
    a. Conc of soln? ____________________
    b. Conc/hr? ____________________
    c. Vol/hr? ____________________

283. Order: Pitocin 40 units in NSS 1000 mL, infuse at 15 mU/min
Available: Oxytocin 10 units/mL
    a. Conc of soln? ____________________
    b. Conc/hr? ____________________
    c. Vol/hr? ____________________

284. Order: Pitocin 10 units in D_5NSS 1000 mL, begin infusion at 1 mU/min
Available: Oxytocin 10 units/mL
    a. Conc of soln? ____________________
    b. Conc/hr? ____________________
    c. Vol/hr? ____________________

285. Order: Pitocin 10 units in D_5NSS 1000 mL, increase infusion to 4 mU/min
Available: Oxytocin 10 units/mL
    a. Conc of soln? ____________________
    b. Conc/hr? ____________________
    c. Vol/hr? ____________________

286. Order: Pitocin 10 units in D_5NSS 1000 mL, increase infusion to 12 mU/min
Available: Oxytocin 10 units/mL
    a. Conc of soln? ____________________
    b. Conc/hr? ____________________
    c. Vol/hr? ____________________

287. Order: Pitocin 10 units in D_5NSS 1000 mL, increase infusion to 16 mU/min
Available: Oxytocin 10 units/mL
    a. Conc of soln? ____________________
    b. Conc/hr? ____________________
    c. Vol/hr? ____________________

For problems 288 and 289, answer these questions:

a. What is the concentration of the solution?
b. What is the concentration per minute?
c. What is the volume per minute?

288. Order: Start pitocin drip, 10 units in NSS 1000 mL, at 12 mL/hr

Available: Oxytocin 10 units/mL

a. Conc of soln? ____________________

b. Conc/min? ____________________

c. Vol/min? ____________________

289. Order: Increase pitocin drip to 60 mL/hr

Available: Pitocin 10 units in NSS 1000 mL

a. Conc of soln? ____________________

b. Conc/min? ____________________

c. Vol/min? ____________________

BODY SURFACE AREA (BSA) NOMOGRAM FOR ADULTS

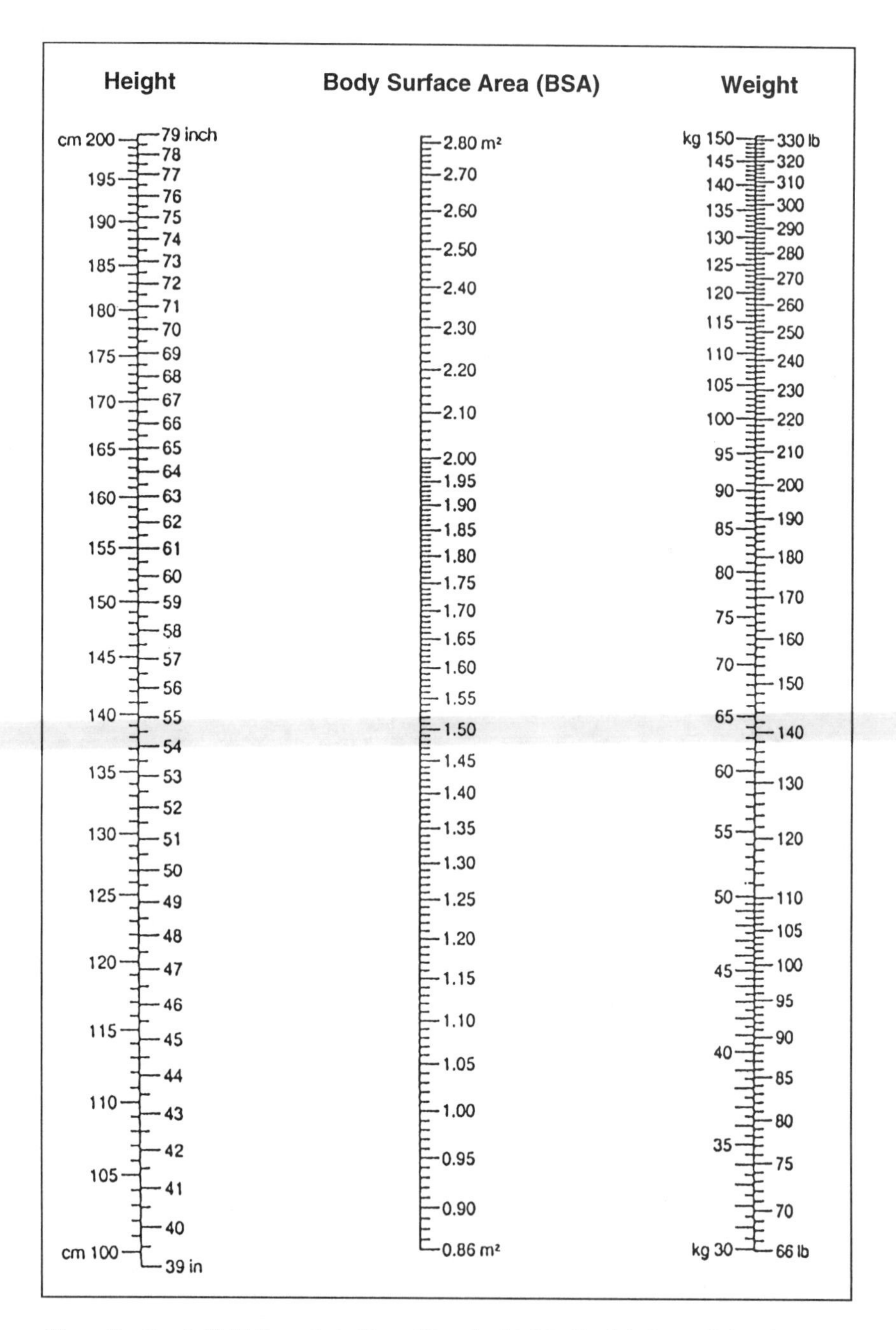

(From Deglin, J. H. Vallerand, A. H. and Russin, M. M.: Davis's Drug Guide for Nurses, 2nd ed. Philadelphia, F. A. Davis Co., 1991, p. 1218. Used with permission from C. Lentner (Ed.), Geigy Scientific Tables, 8th ed., Vol. 1, Ciba-Geigy, Basel, Switzerland, pp. 226–227, 1981.)

BODY SURFACE AREA (BSA) NOMOGRAM FOR CHILDREN

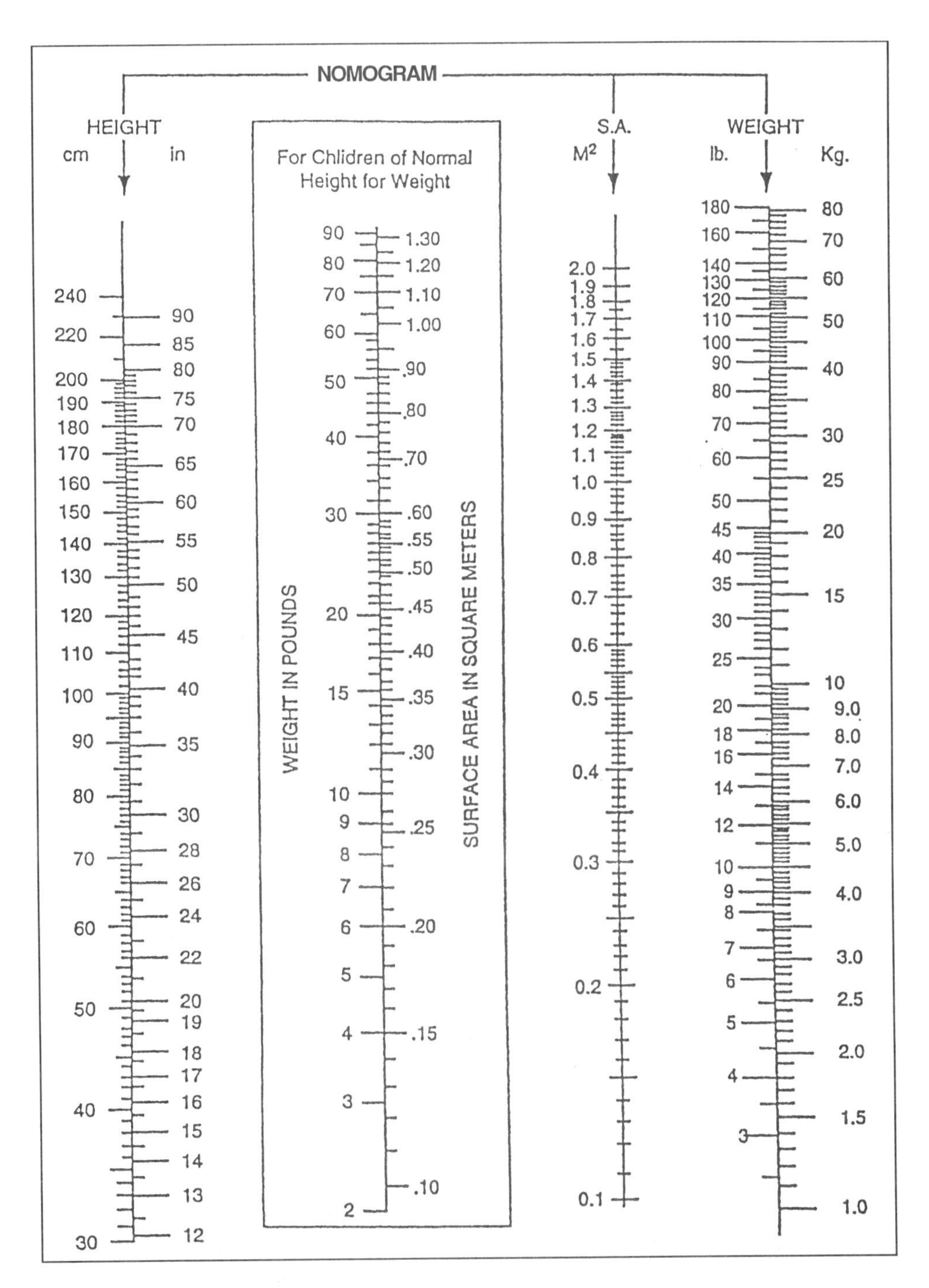

West Nomogram (From Behrman, R. E, and Vaughan, V. C.: Nelson Textbook of Pediatrics, 14th ed. Philadelphia, W. B. Saunders, 1992.)

PART FOUR

Testbank Answers

ORALS (TABLETS AND CAPSULES)

1. a. HydroDiuril 25 mg (You could use HydroDiuril 100 mg bottle, but the tablet should be scored and broken in half.)
   b. $\frac{D}{H} \times V$ $= \frac{50}{25} \times 1$ = 2 tablets
      H : V :: D : x
      25 : 1 :: 50 : x
      $25x = 50$
      x = 2 tablets

2. a. Neither of the bottles
   b. None
   c. Capsules should not be broken in half. Use liquid drug.

3. a. Gr X = 650 mg
   b. 2 tablets of aspirin per dose

4. a. Either bottle; 0.5 g = 500 mg
   b. 2 tablets of Augmentin 250 mg; 1 tablet of Augmentin 500 mg

5. a. Sinemet 25–250 mg bottle
   b. $\frac{D}{H} \times V$ $= \frac{12.5/125}{25/250} \times 1$ = 1/2 tablet
      H : V :: D :x
      25/2 : 1 :: 12.5/125:x
      $25/250x = 12.5/125$
      $x = \frac{12.5/125}{25/250}$
      x = 1/2 tablet

6. 2 capsules

7. a. Nitrostat 0.3 mg. Sublingual (SL) tablet cannot be broken.
   b. 2 tablets SL

8. a. None from this bottle
   b. SL tablet cannot be broken.

9. a. Haldol bottles 2 mg and 1 mg
   b. 1 tablet from each bottle of Haldol 2 mg and 1 mg

10. a. 500 mg
    b. 2 tablets

11. a. 1 g = 1000 mg; 1000 mg ÷ 4 = 250 mg, q6h
    b. 1 capsule of Cloxacillin per dose

12. a. Cardizem SR
    b. 240 mg per day (b.i.d. means twice a day)
    c. 1 capsule of Cardizem SR per dose

13. a. 1.5 mg of Decadron per day (t.i.d. means three times per day)
    b. $\frac{D}{H} \times V$ $= \frac{0.5}{0.25} \times 1$ $= 0.25\overline{)0.50}$ = 2. = 2 tablets
       H : V :: D : x
       0.25 : 1 :: 0.5 : x
       $0.25x = 0.5$
       x = 2 tablets

14. 3 tablets

15. a. 2 g per day
    b. 1 tablet of Procan SR per dose

16. 2 tablets of Mandelamine 0.5 g

17. a. 1 g = 1000 mg; 1.5 g per day
    b. 1500 mg per day
    c. 3 tablets of methydopa per dose.

18. a. Lanoxin 250 μg (0.25 mg) bottle
    b. 2 tablets (or 4 tablets of 125 μg bottle)

19. a. Parameter: 15 mg × 68 kg = 1020 mg per day. Ordered dose is 1 g or 1000 mg per day, which is within safe dose parameters.
    b. 0.5 g = 500 mg; 1 capsule per dose

20. a. 0.25 g = 250 mg (0.250 mg)
    b. 2 tablets of cefuroxime 125 mg per dose.

21. a. 1.2 g = 1200 mg; 1200 ÷ 3 = 400 mg per dose
    b. 2 capsules of Etodolac 200 mg per dose

22. a. Yes; 10 mg, q.i.d. = 40 mg/day (q.i.d. means 4 times a day)
    b. 2 tablets of Compazine per dose

23. a. Either bottle of Glyburide is O.K., but 1.25 mg bottle is preferred because it prevents breaking a 5-mg tablet in half.
    b. 2 tablets of Glyburide 1.25 mg, or 1/2 tablet of Glyburide 5 mg

24. a. Prazosin 2 mg bottle
    b. 2 capsules per dose of Prazosin 2 mg

25. a. Nitrostat 0.6 mg
    b. SL (under the tongue). Drug is absorbed by the sublingual vessels under the tongue because gastric juices destroy the drug.

26. a. 3 mg of Ativan per day
    b. 2 tablets per dose of Ativan 0.5 mg

27. a. 110 pounds ÷ 2.2 kg = 50 kg
    b. 15 mg × 50 kg = 750 mg per day. Dose is within safe dose parameter of 800 mg per day.
    c. Ethambutol 400 mg bottle
    d. 2 tablets from Ethambutol 400 mg bottle per day (or 8 tablets from Ethambutol 100 mg bottle)

28. a. 1.5 g = 1500 mg (1.500 mg)
    b. 3 tablets of Aminocaproic acid from Amicar 500 mg bottle

29. a. Cipro 250 mg bottle
    b. 1 tablet from Cipro 250 mg bottle

30. a. 1gr = 60 mg (see conversion table)
    b. 2 tablets from codeine 30 mg bottle

31. a. Propranolol 10 mg and 20 mg bottles
    b. 1 tablet from each bottle per dose

32. a. 0.4 g = 400 mg; either bottle is O.K., but Tagamet 400 mg is preferred.
    b. Using Tagamet 400 mg bottle, give 1 tablet per dose during the day and 2 tablets at night (h.s.). Using Tagamet 200 mg bottle, give 2 tablets per dose during the day and 4 tablets at night.

33. a. two divided doses q12h
    b. 0.5 g or 500 mg per dose
    c. Either bottle is O.K.,but Cinoxacin 500 mg is preferred.
    d. Using the Cinoxacin 500 mg, give 1 capsule per dose; using Cinoxacin 250 mg, give 2 capsules per dose.

34. a. Captopril 25 mg bottle
    b. 2 tablets of captopril per dose

35. a. $\frac{D}{H} \times V$

$= \frac{2.4}{0.6} \times 1$

$= 4$ tablets

$H : V :: D : x$

$0.6 : 1 :: 2.4 : x$

$0.6x = 2.4$

$x = \frac{2.4}{0.6}$

$x = 4$ tablets

36. a. Decadron 1.5 mg bottle
    b. 2 tablets per dose (or if the 6.0 mg tablet is scored, you can break the tablet in half and give 1/2 tablet)

37. a. 0.1 g = 100 mg (0.100 mg); phenytoin 100 mg bottle
    b. 1 capsule from phenytoin 100 mg bottle

38. a. 8 hours (q8h)
    b. Per dose: (1) 300 mg, (2) 300 mg, and (3) 400 mg
    c. Yes, the dose is within safe dose parameters (900–1080 mg).
       Parameters: 15 mg × 60 = 900 mg;
       18 mg × 60 = 1080 mg
    d. (1) 3 capsules, (2) 3 capsules, and (3) 4 capsules

39. a. 1.2 g = 1200 mg; 1200 ÷ 3 = 400 mg per dose
    b. Either bottle is O.K., but meprobamate 400 mg bottle is preferred.
    c. 1 tablet per dose from meprobamate 400 mg bottle

40. a. 100 pounds ÷ 2.2 = 45.5 kg
    b. 1 mg × 45.5 = 45.5 mg;
       5 mg × 45.5 = 227.3 mg
       Yes, dose is within safe parameters.
    c. 4 tablets of Cytoxan per dose

ORAL SUSPENSION

41. a. $\frac{D}{H} \times V$ H : V :: D : x

$= \frac{100}{250} \times 5$ mL 250 : 5 :: 100 : x

$250x = 500$

$x = 2$ mL

$= \frac{500}{250} = 2$ mL of Dilantin per dose

42. a. 4 mL of Amoxicillin per dose

43. a. $\frac{D}{H} \times V$ H : V :: D : x

$= \frac{1}{2} \times 5$ mL 2 : 5 :: 1 : x

$2x = 5$

$x = 2.5$ mL

$= \frac{5}{2} = 2.5$ mL of Artane per dose

44. a. $\frac{D}{H} \times V = \frac{5}{2} \times 5 = \frac{25}{2} = 12.5$ mL of Artane per dose

45. a. 10 mL of minocycline per dose

46. a. $\frac{D}{H} \times V = \frac{150}{125} \times 5 = \frac{30}{5} = 6$ mL of Ampicillin per dose

47. a. 500 mg of tetracycline per dose
    b. 20 mL of tetracycline per dose

48. a. $\frac{D}{H} \times V = \frac{30}{20} \times 15 = \frac{45}{2} = 22.5$ mL of potassium chloride per dose

49. a. 20 mL of Colace per day

50. a. 20 mg/kg/d × 60 kg = 1200 mg/d
    b. Yes, maximum dose is 1500 mg/d.
    c. $\frac{D}{H} \times V = \frac{1200}{250} \times 5 = \frac{120}{5} = 24$ mL per day

51. a. 7.5 mL of amantadine per dose

52. a. 500 mg per dose
    b. Either bottle is O.K., but Duricef 500 mg/5 mL bottle is preferred.
    c. Using Duricef 500 mg/5 mL bottle, give 5 mL per dose. Using Duricef 250 mg/5 mL, give 10 mL per dose.

53. a. 0.5 g = 500 mg; thus, give 10 mL of cephalexin per dose

54. a. 7.5 mL of Compazine per dose

55. a. $\frac{D}{H} \times V = \frac{25}{10} \times 5 = \frac{25}{2} = 12.5$ mL of Thorazone per dose

56. a. Give 22.5 mL of Theophylline per day.

57. a. Loading dose is 275 mg according to weight, so give 27.5 mL of Theophylline as a loading dose.

58. a. 0.25 g = 250 mg; thus, give 10 mL of Augmentin per dose

59. a. 0.5 g = 500 mg; thus, give 10 mL of ampicillin per dose

60. a. 8 mL of Aldomet per dose

61. a. $\frac{D}{H} \times V = \frac{4}{2} \times 5 = \frac{20}{2} = 10$ mL of albuterol per dose

62. a. $\frac{D}{H} \times V = \frac{200}{300} \times 5 = \frac{10}{3} = 3.3$ mL of Tagamet per dose

63. a. 22.5 mL of potassium chloride per dose
    b. 30 mEq of KCl per day
    45 mL of KCl per day

64. a. 3.75 mL of furosemide per dose

65. a. 10 mL of Zovirax per dose

66. a. 20 mL of diphenydramine per dose

67. a. 45 mL of lactulose per dose

68. a. 20 mL of guaifenesin per dose

69. a. $\frac{D}{H} \times V = \frac{25}{6.25} \times 5 = \frac{125}{6.25} = 20$ mL of promethazine per dose

70. a. 6 hours (q6h)
    b. 2 g = 2000 mg ÷ 4 = 500 mg per dose
    c. 10 mL of erythromycin per dose

71. a. 4 mL of clindamycin per dose

72. a. 0.5 million units = 500,000 units; thus, give 5 mL of Mycostatin per dose

INJECTABLES

73. a. Heparin 5,000 unit cartridge
    b. 0.25 mL of heparin per dose

74. a. Heparin 10,000 unit cartridge
    b. 0.8 mL of heparin per dose

75. a. Tuberculin syringe
    b. S.C. (subcutaneously)
    c. 0.75 mL of heparin per dose.

76. a. 12 U

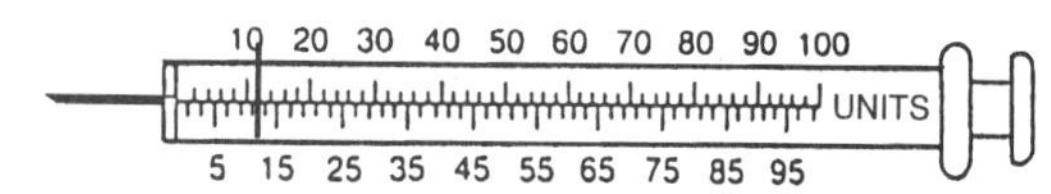

77. a. 25 U

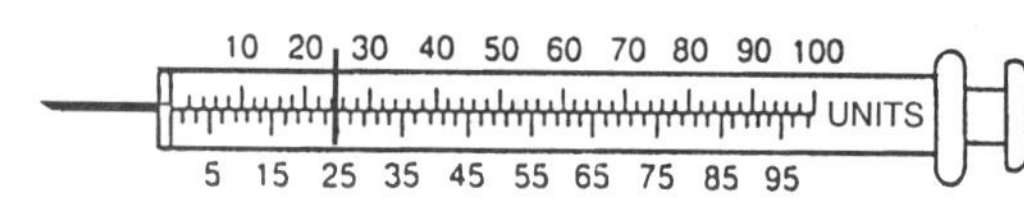

78. a. 42 U

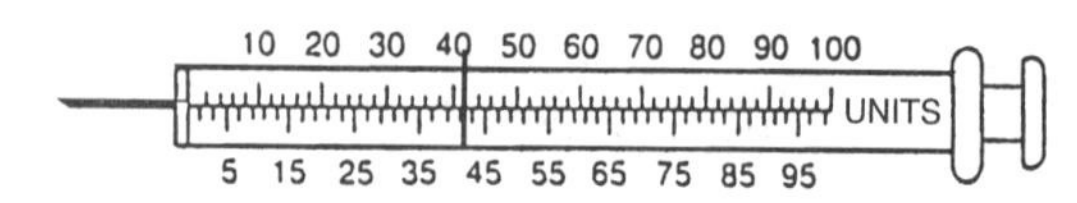

79. a. Inject 5 units of air into the Regular insulin bottle and 45 units of air into the Lente insulin bottle. Withdraw 5 units of Regular insulin and then 45 units of Lente insulin, for a total amount of 50 units of insulin.
    b.

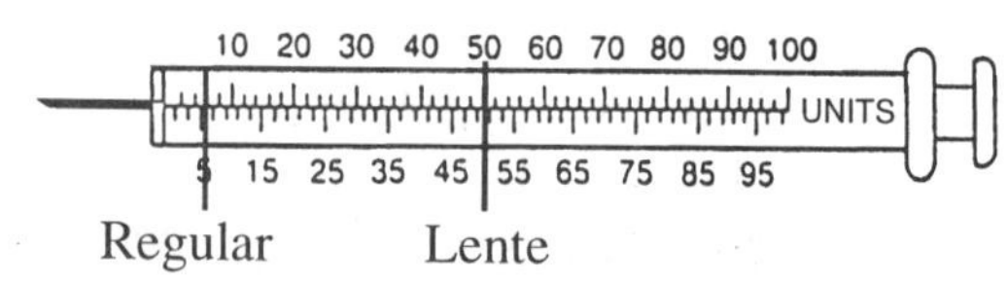

80. a. Inject 6 units of air into the Regular insulin bottle and 40 units of air into the NPH insulin bottle. Withdraw 6 units of Regular insulin and then 40 units of NPH insulin, for a total amount of 46 units.
    b.

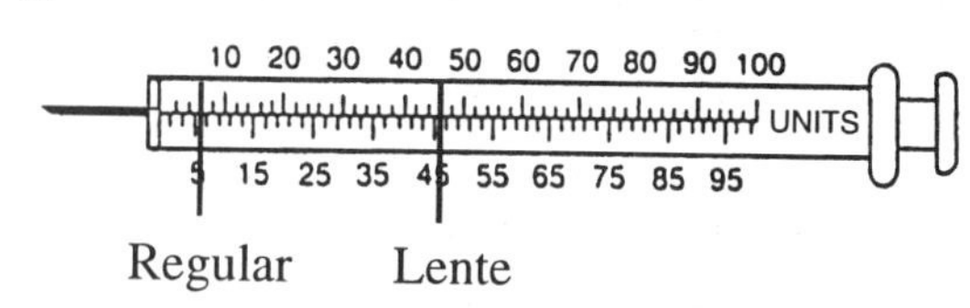

81. a. 10 U Regular; 36 U NPH

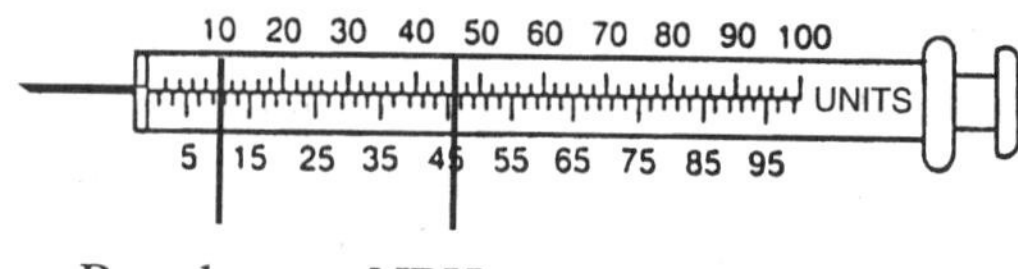

Regular NPH

82. a. Give 0.5 mL of Narcan.
    b. No, an ampul of drug cannot be saved or refrigerated.

83. a. 1/6 gr = 10 mg; thus, give 0.67 mL of morphine

84. a. Cyanocobalamin 1,000 mcg/mL
    b. 0.5 mL of the drug

85. a. $\frac{D}{H} \times V = \frac{\overset{3}{\cancel{75}}}{\underset{4}{\cancel{100}}} \times 2 = \frac{3}{2} = 1.5$ mL of hydroxyzine

86. a. Meperidine 30 mg = 0.6 mL
    Atropine 0.3 mg = 0.75 mL
    (total amount = 1.35 mL)
    b. From the meperidine cartridge, expel 0.4 mL of the drug (0.6 mL should remain). Draw 0.8 mL of air into the cartridge and inject into the atropine vial. Withdraw 0.75 mL of atropine into the cartridge.

87. a. Meperidine:

$$\frac{D}{H} \times V = \frac{70}{100} \times 1 = 0.7 \text{ mL}$$

Atropine:

$$H : V :: D : x$$
$$0.4 : 1 :: 0.6 : x$$
$$0.4x = 0.6$$
$$x = \frac{0.6}{0.4} = 1.5 \text{ mL}$$

(total amount = 2.2 mL)

    b. From the meperidine cartridge, expel 0.3 mL of the drug (0.7 mL should remain). Draw 1.5 mL of air into the cartridge and inject into the atropine vial. Withdraw 1.5 mL of atropine into the cartridge.

88. a. $15 \text{ mg} \times 90 \text{ kg} = 1350 \text{ mg}$
 b. 1200 mg of Kantrex per day
 c. Yes, dose per day (1200 mg) is less than maximum dose (1350 mg).
 d. 1 g = 1000 mg; thus,
 $\frac{D}{H} \times V = \frac{400}{1000} \times 3 = \frac{12}{10} = 1.2$ mL of Kantrex

89. a. 67.7 kg
 b. $15 \times 67.7 = 1015.5$ mg
 c. Yes, the dose per day (1000 mg) is within safe parameters.
 d. 1 g = 1000 mg, and 1000 mg = 4 mL of Amikin; thus, give 2 mL of Amikacin per dose

90. a. 0.8 mL of Thorazine deep IM per dose

91. a. $0.67 \approx 0.7$ mL of Tagamet per dose

92. a. 1.5 mL of Compazine deep IM stat

93. a. 0.5 mL of Compazine per dose

94. a. 2 mg of Haldol per dose

95. a. 2 mL of diluent
 b. 2.6 mL (= 1 g)
 c. 1 g = 1000 mg (or 0.5 g = 500 mg); thus
 $\frac{D}{H} \times V = \frac{500}{1000} \times 2.6 = \frac{2.6}{2} = 1.3$ mL of Ticar per dose

96. a. 2 mL of diluent
 b. 2.2 mL (= 500 mg)
 c. 0.25 g = 250 mg; thus, give 1.1 mL of Ancef per dose

97. a. 5.7 mL of diluent (1 g = 6 mL)
 b. 1.5 mL of oxacillin per dose

98. a. 6.6 mL of diluent
 b. 8 mL (= 2 g)
 c. $\frac{D}{H} \times V = \frac{0.5}{2} \times 8 = \frac{4}{2} = 2$ mL of nafcillin per dose

99. a. 0.2 g = 200 mg; thus,
 $\frac{D}{H} \times V = \frac{200}{300} \times 2 = \frac{4}{3} = 1.3$ mL of Tagamet per dose

100. a. 0.5 g = 500 mg; thus, give 2 mL of Dyphylline per dose

101. a. $1.67 \approx 1.7$ mL of Clindamycin per dose

102. a. 3.5 mL of diluent
 b. 4 mL
 c. 1 g = 1000 mg; thus, give 2 mL of ampicillin per dose

103. a. 3 mL of diluent
 b. 3.5 mL
 c. 1 g = 1000 mg; thus
 H : V :: D : x
 1000 : 3.5 :: 750 : x
 $1000x = 2625$
 $x = \frac{2625}{1000} = 2.6$ mL of Mandol per dose

104. a. 0.75 mL of morphine prn

105. a. 2.0 mL of diluent
 b. 2.4 mL
 c. $\frac{D}{H} \times V = \frac{600}{1000} \times 2.4 = \frac{14.4}{10} = 1.4$ mL of Cefadyl per dose

106. 0.75 mL of Decadron stat

107. a. 1/100 gr = 0.6 mg; thus, give 1.5 mL of Atropine stat

108. a. $\frac{D}{H} \times V = \frac{3}{10} \times 10 = 3$ mL of Moxam per dose

109. a. $3 \text{ mg} \times 64 = 192$ mg/d (or 64 mg per dose, and
 $5 \text{ mg} \times 64 = 320$ mg/d (or 107 mg per dose;
 thus, the dosage range is 192–320 mg/d
 b. 64–107 mg per dose
 c. Using 64 mg per dose,
 H : V :: D : x
 80 : 2 :: 64 : x
 $80x = 128$
 $x = 1.6$ mL of tobramycin per dose

110. a. 1.75 mL of Azactam per dose

111. a. 0.7 mL of Demerol per dose

PEDIATRICS

112. a. Yes. The dose is 80 mg × 2/d (q12h) = 160 mg/d, and parameter is 8 mg/kg/d × 20 kg = 160 mg/d, or the same.
 b. 4 mL of Suprax per dose

113. a. 75 mg × 3/d (q8h) = 225 mg/d
 b. Yes. The dose (20 mg/kg/d × 32 kg = 640 mg/d) is less than the parameter (40 mg/kg/d × 32 kg = 1280 mg/d).
 c. 3 mL of Amoxil per dose (The nurse may want to check with the physician to see if this dose is adequate.)

114. a. 150 mg × 4/d (q6h) = 600 mg/d
 b. **No.** The clindamycin dose (600 mg/d) exceeds the safe dose parameters (8 mg/kg/d × 18 kg = 144 mg/d to 25 mg/kg/d × 18 kg = 450 mg/d). Drug dose per day is unsafe.
 c. **Do not give;** notify the physician at once.

115. a. 66 lbs ÷ 2.2 kg/lb = 30 kg
 b. Order: 250 mg × 3 (q8h) = 750 mg/d
 c. Yes. The dose per day (750 mg/d) is within safe dose parameters (20 mg/kg/d × 30 kg = 600 mg/d to 40 mg/kg/d × 30 kg = 1200 mg/d).
 d. 5 mL of Ceclor per dose

116. a. 30 lbs ÷ 2.2 kg/lb = 13.6 kg
 b. 300 mg/d
 c. Yes. The dose per day (300 mg/d) is within safe dose parameters (20 mg/kg/d × 13.6 kg = 272 mg/d to 40 mg/kg/d × 13.6 kg = 544 mg/d).
 d. 4 mL of Ceclor per dose

117. a. 400 mg/d
 b. Yes. The dose per day (400 mg/d) is within safe parameters (15 mg/kg/d × 14 kg = 210 mg/d to 50 mg × 14 kg = 700 mg/d).
 c. 4 mL of Penicillin V per dose

118. a. 100 mg/d
 b. Parameter: 3 mg/kg/d × 16 kg = 48 mg/d
 8 mg/kg/d × 16 kg = 128 mg/d
 (48–128 mg/d)
 c. Yes
 d. 2.5 mL of phenytoin per dose

119. a. BSA = 0.74 m^2
 b. 250 $mg/m^2/d$ × 0.74 m^2 = 185 mg/d
 c. Yes. The dose per day (125 mg/d) is less than maximum dose (185 mg/d).
 d. Dilantin 100 mg bottle
 e. 6.25 mL from Dilantin 100 mg/5 mL bottle

120. a. Parameter: 30 mg/kg/d × 32 kg = 960 mg/d
 b. Yes. The drug dose (800 mg/d) is less than the maximum dose (960 mg/d).
 c. Either bottle is O.K., but Duricef 250 mg/5 mL is preferred.
 d. $\frac{D}{H} \times V = \frac{400}{250} \times 5$ = 8 mL of Duricef per dose (handwritten: 400 → 8, 250 → 5)

121. a. 600 mg/d
 b. Yes. The drug dose (600 mg/d) is within safe parameters (50 mg/kg/d × 12 kg = 600 mg/d to 100 mg/kg/d × 12 kg = 1200 mg/d).
 c. 3 mL of ampicillin per dose

122. a. 110 lbs ÷ 2.2 kg/lb = 50 kg
 b. 500 mg of Aldomet per day
 c. Yes. The dose (500 mg/d) is within safe dose parameters (10 mg/kg/d × 50 kg = 500 mg/d).
 d. 5 mL of Aldomet per dose

123. a. BSA = 0.84 m^2
 b. Parameter: 100 $mg/m^2/d$ × 0.84 m^2 = 84 mg/d
 c. 1/4 gr = 15 mg. Order is for 60–90 mg/d. But since maximum dose is 84 mg/d, give no more than 5 times per day. (15 mg × 5/d = 75 mg/d). **Not safe** for 6 times per day (15 mg × 6/d = 90 mg/d).
 d. 1 tablet not more than 5 times a day, prn

124. a. 8 lbs ÷ 2.2 kg/lb = 3.6 kg
Parameters:
0.035 mg/kg/d × 3.6 kg = 0.126 mg/d
0.060 mg/kg/d × 3.6 kg = 0.216 mg/d
b. **No.** The order is for 0.1 mg × 3/d (q8h) = 0.3 mg/d, which is more than the safe dose parameter (0.216 mg/d).
c. **Do not give;** notify the physician at once.

125. a. 8 lbs ÷ 2.2 kg/lb = 3.6 kg
Parameters:
0.01 mg/kg/d × 3.6 kg = 0.036 mg/d
0.02 mg/kg/d × 3.6 kg = 0.072 mg/d
b. **No.** The order is for 0.05 mg × 2/d (q12h) =0.1 mg/d, which is more than the safe dose parameter (0.072 mg/d).
c. **Do not give;** notify the physician at once.

126. a. 33 lbs ÷ 2.2 kg/lb = 15 kg
b. Parameters:
120 mg/kg/d × 15 kg = 1800 mg/d
150 mg/kg/d × 15 kg = 2250 mg/d
c. Yes. The order is for 300 mg × 4/d (q6h) = 1200 mg/d, which is less than the safe dose parameters (1800–2250 mg/d).
d. 3.0 mL of Gantrisin per dose. The nurse may want to check with the physician to see if this dose is adequate.

127. a. BSA = 1.16 m^2
b. 4 $g/m^2/d \times 1.16\ m^2$ = 4.64 g/d
c. Yes. The order is for 0.75 g × 4/d (q6h) = 3 g/d, which is less than the safe dose parameter (4.64 g/d).
d. 0.75 g = 750 mg per dose
$\frac{D}{H} \times V = \frac{\cancel{750}\ 3}{\cancel{500}\ 2} \times 5 = 7.5$ mL of Gantrison per dose

128. a. Parameters:
TMP 8 mg/kg/d × 15 kg = 120 mg/d
SMZ 40 mg/kg/d × 15 kg = 600 mg/d
b. Yes. The order is for 60/300 mg × 2/d (q12h) = 120/600 mg/d, which is within safe dose parameters (120/600 mg/d).
c. 7.5 mL of Septra per dose

129. a. 60 lbs ÷ 2.2 kg/lb = 27.27 kg
b. Yes. The order is for 0.25 g × 3/d (q8h) = 0.75 mg/d, or 750 mg/d, which is within safe dose parameters:
20 mg/kg/d × 27.27 kg = 545 mg/d to
40 mg/kg/d × 27.27 kg = 1090 mg/d
c. 5 mL of Augmentin per dose

130. a. BSA = 1.3 m^2
b. Yes. The parameter and the prescribed dose are the same:
4.5 mg/m^2/dose × 1.3 m^2 = 5.85 mg/dose or 6 mg per single dose
c. Three Leukeran 2 mg tablets per dose

131. a. 22 lbs ÷ 2.2 kg/lb = 10 kg
b. **No.** Prescribed dose (25 mg × 4/d (q6h) = 100 mg/d) exceeds safe dose parameter (5 mg/kg/d × 10 kg = 50 mg/d).
c. **Do not give.** Notify the physician at once. Prescribed dose can be cut in half if ordered by the physician.

132. a. Yes. The order is for 120 mg/dose × 4 doses per day = 480 mg/d, which is within the safe dose parameters (120–160 mg per dose; not to exceed 480 mg/d).
b. 80 mg = 2.5 mL (1/2 teaspoon)
$\frac{D}{H} \times V = \frac{\cancel{120}\ 3}{\cancel{80}\ 2} \times 2.5 = \frac{7.5}{2} = 3.75$ mL of Tylenol per dose

133. a. BSA = 0.92 m^2
b. Parameter: 500 mg/m^2/dose × 0.92 m^2 = 460 mg/dose
460 mg/dose × 3 doses/d (q8h) = 1380 mg/d
c. Yes. The dose is 400 mg (q8h) = 1200 mg/d, which is within safe dose parameters (1380 mg/d).
d. $\frac{D}{H} \times V = \frac{\cancel{400}\ 4}{\cancel{200}\ 5} \times 5 =$ 10 mL of Zovirax per dose

134. a. Yes. The dose is 250 mg/d, which is less than the safe dose parameter (20 mg/kg/d × 24 kg = 480 mg/d).
b. 2.5 mL of Zarontin per dose (or 2/day)

135. a. Ritalin 5 mg bottle.
b. 1 1/2 tablets of Ritalin every morning

136. a. Use both Ritalin 5 mg and 10 mg bottles, or use only the Ritalin 5 mg bottle.
  b. If both bottles are used, give 1 tablet from Ritalin 5 mg bottle and 1 tablet from Ritalin 10 mg bottle. If only Ritalin 5 mg bottle is used, give 3 tablets.

137. a. The order is for 15 mg × 4/d (q6h) = 60 mg/d.
  b. Parameters:
  6 mg/kg/d × 12 kg = 72 mg/d
  7.5 mg/kg/d × 12 kg = 90 mg/d
  c. Yes. The dose (60 mg/d) is less than safe dose parameters (72–90 mg/d).
  d. Tobramycin 15 mg, IV = 1.5 mL

138. a. 44 lbs ÷ 2.2 kg/lb = 20 kg
  b. Parameter: 15 mg/kg/d × 20 kg = 300 mg/d
  c. Yes. The dose is 50 mg × 3/d (q8h) = 150 mg/d, which is less than the safe dose parameter (300 mg/d).
  d. H : V :: D : x
  75 : 2 :: 50 : x

$$75x = 100$$
$$x = \frac{100}{75} = 1.3 \text{ mL per dose}$$

139. a. Lanoxin 0.5 mg bottle
  b. Parameters:
  0.015 mg/kg × 16 kg = 0.24 mg
  0.035 mg/kg × 16 kg = 0.56 mg
  c. Yes. The dose is 0.25 mg, which is within the dose parameters (0.24–0.56 mg).
  d. 0.25 mg = 1 mL, IV

140. a. Parameters:
  50 mg/kg/d × 13 kg = 650 mg/d
  100 mg/kg/d × 13 kg = 1300 mg/d
  b. Yes. The dose is 200 mg × 4/d (q6h) = 800 mg/d, which is within safe dose parameters (650–1300 mg/d).
  c. 1.4 mL of sterile water (after reconstitution, drug solution = 1.5 mL)

$$\frac{D}{H} \times V = \frac{200}{250} \times 1.5 = 1.2 \text{ mL of Oxacillin per dose}$$

(handwritten: 200 → 4, 250 → 5)

141. a. 0.3 mL of Narcan

DRUG CALCULATION BY BODY SURFACE AREA (BSA)

142. Height 70 in., weight 173 lbs, intersects 2.02 m^2
$250 \text{ mg/m}^2\text{/d} \times 2.02 \text{ m}^2 = 505 \approx 500 \text{ mg/d}$

143. Height 66 in., weight 210 lbs, intersects 2.04 m^2
$450 \text{ mg/m}^2\text{/wk} \times 2.04 \text{ m}^2 = 918 \approx 920 \text{ mg/wk}$

144. Height 66 in., weight 210 lbs, intersects 2.04 m^2
$200 \text{ mg/m}^2\text{/wk} \times 2.04 \text{ m}^2 = 408.00 \approx 400 \text{ mg/wk}$

145. Height 72 in., weight 80 kg, intersects 2.08 m^2
$30 \text{ mg/m}^2\text{/d} \times 2.08 \text{ m}^2 = 62.4 \approx 62 \text{ mg/d}$

146. Height 74 in., weight 186 lbs, intersects 2.10 m^2
$80 \text{ mg/m}^2\text{/wk} \times 2.10 \text{ m}^2 = 168 \approx 170 \text{ mg/wk}$

147. Height 74 in., weight 70 kg, intersects 2.06 m^2
$120 \text{ mg/m}^2\text{/wk} \times 2.06 \text{ m}^2 = 247.2 \approx 250 \text{ mg/wk}$

148. Height 70 in., weight 85 kg, intersects 2.08 m^2
$600 \text{ mg/m}^2\text{/wk} \times 2.08 \text{ m}^2 = 1248 \approx 1250 \text{ mg/wk}$

149. Height 70 in., weight 85 kg, intersects 2.08 m^2
$60 \text{ mg/m}^2\text{/wk} \times 2.08 \text{ m}^2 = 124.8 \approx 125 \text{ mg/wk}$

150. Height 62 in., weight 75 kg, intersects 1.78 m^2
$2 \text{ mg/m}^2\text{/wk} \times 1.78 \text{ m}^2 = 3.56 \approx 3.6 \text{ mg/wk}$

151. Height 65 in.,weight 64 kg, intersects 1.75 m^2
$15 \text{ mg/m}^2\text{/wk} \times 1.75 \text{ m}^2 = 26.25 \approx 26 \text{ mg/wk}$

152. Height 68 in., weight 150 lbs,
    intersects 1.85 m^2
    $12 \text{ mg/m}^2\text{/d} \times 1.85 \text{ m}^2 = 22.2 \approx 22 \text{ mg/d}$

153. Height 62 in., weight 130 lbs,
    intersects 1.65 m^2
    $100 \text{ mg/m}^2\text{/d} \times 1.65 \text{ m}^2 = 165 \text{mg/d}$

154. Height 72 in., weight 82 kg,
    intersects 2.10 m^2
    $3.3 \text{ mg/m}^2\text{/d} \times 2.10 \text{ m}^2 = 6.93 \approx 6.9 \text{ mg/d}$

155. Height 72 in., weight 82 kg,
    intersects 2.08 m^2
    $60 \text{ mg/m}^2\text{/d} \times 2.08 \text{ m}^2 = 124.8 \approx 125 \text{ mg/d}$

156. Height 60 in., weight 60 kg,
    intersects 1.60 m^2
    $12 \text{ mg/m}^2\text{/d} \times 1.60 \text{ m}^2 = 19.2 \approx 19 \text{ mg/d}$

157. Height 64 in., weight 60 kg,
    intersects 1.69 m^2
    $100 \text{ mg/m}^2\text{/d} \times 1.69 \text{ m}^2 = 169 \approx 170 \text{ mg/d}$

INTRAVENOUS FLUIDS

Hours to Administer

158. $\frac{\text{Volume of soln}}{\text{Rate}} = \frac{1000 \text{ mL}}{30 \text{ mL/hr}} = 33.3 \text{ hr}$

159. $\frac{\text{Volume of soln}}{\text{Rate}} = \frac{1000 \text{ mL}}{100 \text{ mL/hr}} = 10 \text{ hr}$

160. $\frac{\text{Volume of soln}}{\text{Rate}} = \frac{1000 \text{ mL}}{125 \text{ mL/hr}} = 8 \text{ hr}$

161. $\frac{\text{Volume of soln}}{\text{Rate}} = \frac{500 \text{ mL}}{20 \text{ mL/hr}} = 25 \text{ hr}$

162. $\frac{\text{Volume of soln}}{\text{Rate}} = \frac{1000 \text{ mL}}{90 \text{ mL/hr}} = 11.1 \text{ hr}$

163. $\frac{\text{Volume of soln}}{\text{Rate}} = \frac{1000 \text{ mL}}{40 \text{ mL/hr}} = 25 \text{ hr}$

164. $\frac{\text{Volume of soln}}{\text{Rate}} = \frac{1000 \text{ mL}}{50 \text{ mL/hr}} = 20 \text{ hr}$

165. $\frac{\text{Volume of soln}}{\text{Rate}} = \frac{1000 \text{ mL}}{150 \text{ mL/hr}} = 6.6 \text{ hr}$

166. $\frac{\text{Volume of soln}}{\text{Rate}} = \frac{1000 \text{ mL}}{60 \text{ mL/hr}} = 16.6 \text{ hr}$

167. $\frac{\text{Volume of soln}}{\text{Rate}} = \frac{1000 \text{ mL}}{80 \text{ mL/hr}} = 12.5 \text{ hr}$

Direct IV Injection (IV Push)

168. $\frac{D}{H} \times V$
    $= \frac{0.6 \text{ mg}}{1 \text{ mg}} \times 10 \text{ mL} = 6 \text{ mL}$
    or
    $H : V :: D : x$
    $1 \text{ mg} : 10 \text{ mL} :: 0.6 \text{ mg} : x \text{ mL}$
    $x = 10 \times 0.6$
    $x = 6 \text{ mL}$

169. $\frac{80 \text{ mg}}{100 \text{ mg}} \times 10 \text{ mL} = 8 \text{ mL}$

170. $\frac{75 \text{ mg}}{100 \text{ mg}} \times 10 \text{ mL} = 7.5 \text{ mL}$

171. $\frac{0.75 \text{ mg}}{1 \text{ mg}} \times 10 \text{ mL} = 7.5 \text{ mL}$

172. $\frac{0.5 \text{ mg}}{1 \text{ mg}} \times 10 \text{ mL} = 5 \text{ mL}$

173. $\frac{2 \text{ mg}}{5 \text{ mg}} \times 1 \text{ mL} = 0.4 \text{ mL}$

174. $\frac{80 \text{ mg}}{100 \text{ mg}} \times 10 \text{ mL} = 8 \text{ mL}$

175. $\frac{6 \text{ mg}}{3 \text{ mg}} \times 1 \text{ mL} = 2 \text{ mL}$

176. $\frac{12 \text{ mg}}{3 \text{ mg}} \times 1 \text{ mL} = 4 \text{ mL}$

177. $\frac{3 \text{ g}}{1 \text{ g}} \times 1 \text{ mL} = 3 \text{ mL}$

178. $\frac{75 \text{ mg}}{40 \text{ mg}} \times 1 \text{ mL} = 1.87 \approx 1.9 \text{ mL}$

179. $\frac{80 \text{ mg}}{125 \text{ mg}} \times 1 \text{ mL} = 0.64 \approx 0.6 \text{ mL}$

180. a. $0.15 \text{ mg/kg} \times 75 \text{ kg} = 11.25 \text{ mg}$
b. $\frac{11.25 \text{ mg}}{2.5 \text{ mg}} \times 1 \text{ mL} = 4.5 \text{ mL}$

181. $\frac{5 \text{ mg}}{1 \text{ mg}} \times 1 \text{ mL} = 5 \text{ mL}$

182. $\frac{15 \text{ mg}}{20 \text{ mg}} \times 1 \text{ mL} = 0.75 \text{ mL}$

183. $\frac{0.2 \text{ mg}}{0.1 \text{ mg}} \times 1 \text{ mL} = 2 \text{ mL}$

184. a. $1 \text{ mEq/kg} \times 70 \text{ kg} = 70 \text{ mEq}$
b. $\frac{70 \text{ kg}}{44.6 \text{ mEq}} \times 50 \text{ mL} = 78.4 \approx 78 \text{ mL}$

185. $\frac{1 \text{ mg}}{0.5 \text{ mg}} \times 2 \text{ mL} = 4 \text{ mL}$

186. $\frac{800 \text{ mg}}{10 \text{ mg}} \times 1 \text{ mL} = 80 \text{ mL}$

187. $\frac{0.125 \text{ mg}}{0.5 \text{ mg}} \times 2 \text{ mL} = 0.5 \text{ mL}$

188. a. $\frac{900 \text{ mg}}{50 \text{ mg}} \times 1 \text{ mL} = 18 \text{ mL}$
b. $\frac{900 \text{ mg}}{50 \text{ mg}} \times 1 \text{ min} = 18 \text{ min}$ for IV push

189. a. $10 \text{ mg/kg} \times 65 \text{ kg} = 650 \text{ mg}$
b. $\frac{650 \text{ mg}}{30 \text{ mg}} \times 1 \text{ mL} = 21.67 \text{ mL}$

190. a. $0.75 \text{ mg/kg} \times 75 \text{ kg} = 56.25 \approx 56 \text{ mg}$
b. $\frac{56.25 \text{ mg}}{5 \text{ mg}} \times 1 \text{ mL} = 11.25 \approx 11 \text{ mL}$ IV push over 3 min

IV Drop Rate

191. a. $\frac{40 \text{ mL} \times 10 \text{ gtt}}{60 \text{ min}} = 6.6 \text{ gtt/min}$
b. 10 gtt/min
c. 13 gtt/min
d. 40 gtt/min

192. a. $\frac{100 \text{ mL} \times 10 \text{ gtt}}{60 \text{ min}} = 17 \text{ gtt/min}$
b. 25 gtt/min
c. 34 gtt/min
d. 100 gtt/min

193. a. $\frac{80 \text{ mL} \times 10 \text{ gtt}}{60 \text{ min}} = 13 \text{ gtt/min}$
b. 20 gtt/min
c. 26–27 gtt/min
d. 80 gtt/min

194. a. $\frac{125 \text{ mL} \times 10 \text{ gtt}}{60 \text{ min}} = 20.8 \approx 21 \text{ gtt/min}$
b. 31–32 gtt/min
c. 42 gtt/min
d. 125 gtt/min

195. a. $\frac{150 \text{ mL} \times 10 \text{ gtt}}{60 \text{ min}} = 25 \text{ gtt/min}$
b. 38 gtt/min
c. 50 gtt/min
d. 150 gtt/min

196. a. $\frac{200 \text{ mL} \times 10 \text{ gtt}}{60 \text{ min}} = 33 \text{ gtt/min}$
b. 50 gtt/min
c. 66 gtt/min
d. 200 gtt/min

197. a. $\frac{75 \text{ mL} \times 10 \text{ gtt}}{60 \text{ min}} = 12.5 \text{ gtt/min}$
b. 19 gtt/min
c. 25 gtt/min
d. 75 gtt/min

198. a. $\frac{500 \text{ mL} \times 10 \text{ gtt}}{60 \text{ min}} = 83 \text{ gtt/min}$
b. 125 gtt/min (too rapid to count)
c. 166 gtt/min (too rapid to count)
d. 500 gtt/min (too rapid to count)

199. a. $\frac{30 \text{ mL} \times 10 \text{ gtt}}{60 \text{ min}} = 5 \text{ gtt/min}$
b. 8 gtt/min
c. 10 gtt/min
d. 30 gtt/min

200. a. $\frac{60\text{ mL} \times 10\text{ gtt}}{60\text{ min}} = 10\text{ gtt/min}$
   b. 15 gtt/min
   c. 20 gtt/min
   d. 60 gtt/min

201. a. $\frac{150\text{ mL} \times 10\text{ gtt}}{60\text{ min}} = 25\text{ gtt/min}$
   b. 38 gtt/min
   c. 50 gtt/min
   d. 150 gtt/min (too rapid to count)

202. a. $\frac{50\text{ mL} \times 10\text{ gtt}}{60\text{ min}} = 8\text{ gtt/min}$
   b. 13 gtt/min
   c. 16 gtt/min
   d. 50 gtt/min

203. a. $\frac{1000\text{ mL} \times 10\text{ gtt}}{60\text{ min}} = 166\text{ gtt/min}$ (too rapid to count)
   b. 250 gtt/min (too rapid to count)
   c. 333 gtt/min (too rapid to count)
   d. 1000 gtt/min (too rapid to count) Monitor infusion wide open.

204. a. $\frac{20\text{ mL} \times 10\text{ gtt}}{60\text{ min}} = 3.3\text{ gtt/min}$
   b. 5 gtt/min
   c. 7 gtt/min
   d. 20 gtt/min

205. a. $\frac{90\text{ mL} \times 10\text{ gtt}}{60\text{ min}} = 15\text{ gtt/min}$
   b. 23 gtt/min
   c. 30 gtt/min
   d. 90 gtt/min

Antibiotics

206. a. $\frac{D}{H} \times V = \frac{800\text{ mg}}{500\text{ mg}} \times 2\text{ mL} = 3.2\text{ mL}$
   b. $\frac{\text{Volume of soln}}{\text{Time}} \times \text{drop factor}$
   $= \frac{150\text{ mL}}{60\text{ min}} \times \frac{15\text{ gtt}}{1\text{ mL}} = 37.5\text{ gtt/min}$

207. a. 6 mL
   b. 13.8 ≈ 14 gtt/min

208. a. 2.5 mL
   b. 200 gtt/min

209. a. 5 mL
   b. 75 gtt/min

210. a. 5 mL
   b. 150 gtt/min

211. a. 5 mL
   b. 83.3 ≈ 83 gtt/min

212. a 10 mL; 10 mL + 75 mL = 85 mL
   b. 170 gtt/min

213. a. 2 mL
   b. 100 gtt/min

214. a. 6 mL
   b. 33.3 ≈ 33 gtt/min

215. a. 2.5 mL
   b. 100 gtt/min

216. a. 10 mL
   b. 73.3 ≈ 73 gtt/min

217. a. 1.5 mL
   b. 100 gtt/min

218. a. 3 mL
   b. 66.6 ≈ 67 gtt/min

219. a. 1 mL
   b. 100 gtt/min

220. a 3 mL
   b. 50 gtt/min

221. a. 6 mL
   b. 50 gtt/min

222. a. 1.5 mL
   b. 50 gtt/min

223. a. 10 mL; 10 mL + 75 mL = 85 mL
 b. 170 gtt/min

224. a. 1 mL
 b. 33.3 ≈ 33 gtt/min

225. a. 3 mL
 b. 100 gtt/min

226. a. 6.6 mL
 b. 18.8 ≈ 19 gtt/min

227. a. 5 mL
 b. 16.6 ≈ 17 gtt/min

228. a. 15 mL
 b. 21 gtt/min

229. a. 5 mL
 b. 41.6 ≈ 42 gtt/min

230. a. 10 mL; 10 mL + 100 mL = 110 mL
 b. 55 gtt/min

231. a. 10 mL
 b. 170 gtt/min

232. a. 3.75 mL
 b. 100 gtt/min

233. a. 2.5 mL
 b. 100 gtt/min

234. a. 10 mL
 b. 30 gtt/min

235. a. 10 mL; 10 mL + 50 mL = 60 mL
 b. 30 gtt/min

236. a. 1.5 mL
 b. 66.6 ≈ 67 gtt/min

237. a. 4 mL
 b. 83.3 ≈ 83 gtt/min

Critical Care: Infusion Rates

238. a. $\frac{1000 \text{ mg}}{250 \text{ mL}} = 4 \text{ mg/mL}$
 b. $\frac{5 \text{ mL/hr}}{60 \text{ min/hr}} = 0.083 \text{ mL/min}$
 c. $\frac{20 \text{ mg/hr}}{4 \text{ mg/mL}} = 5 \text{ mL/hr}$
 d. $\frac{20 \text{ mg/hr}}{60 \text{ min/hr}} = 0.33 \text{ mg/min}$
 e. 20 mg/hr (given)

239. a. 4 mg/mL
 b. 0.75 mL/min
 c. 45 mL/hr
 d. 3 mg/min (given)
 e. 180 mg/hr

240. a. 1600 mcg/mL (or 1.6 mg/mL)
 b. 0.11 mL/min
 c. 6.75 mL/hr
 d. 3 mcg/kg/min × 60 kg = 180 mcg/min
 e. 10,800 mcg/hr (or 10.8 mg/hr)

241. a. 200 mcg/mL (0.2 mg/mL)
 b. 0.05 mL/min
 c. 3 mL/hr
 d. 10 mcg/min (given)
 e. 600 mcg/hr

242. a. 400 mcg/mL (0.4 mg/mL)
 b. 0.0625 mL/min
 c. 3.75 mL/hr
 d. 25 mcg/min (given)
 e. 1500 mcg/hr (or 1.5 mg/hr)

243. a. 2 mg/mL
 b. 2 mL/min
 c. 120 mL/hr
 d. 4 mg/min (given)
 e. 240 mg/hr

244. a. 500 mcg/mL (0.5 mg/mL)
 b. 0.6 mL/min
 c. 36 mL/hr
 d. 5 mcg/kg/min × 60 kg = 300 mcg/min
 e. 18,000 mcg/hr (or 18 mg/hr)

245. a. 2 mcg/mL (0.002 mg/mL)
   b. 2.5 mL/min
   c. 150 mL/hr
   d. 5 mcg/min (given)
   e. 300 mcg/hr

246. a. 4 mcg/mL (0.004 mg/mL)
   b. 1.25 mL/min
   c. 75 mL/hr
   d. 5 mcg/min (given)
   e. 300 mcg/hr

247. a. 4 mcg/mL (0.004 mg/mL)
   b. 2.5 mL/min
   c. 150 mL/hr
   d. 10 mcg/min (given)
   e. 60 mcg/hr

248. a. 8 mcg/mL (0.008 mg/mL)
   b. 1 mL/min
   c. 60 mL/hr
   d. 8 mcg/min (given)
   e. 480 mcg/hr

249. a. 20 mcg/mL (0.02 mg/mL)
   b. 0.2 mL/min
   c. 12 mL/hr
   d. 4 mcg/min (given)
   e. 240 mcg/hr

250. a. 4 mg/mL
   b. 0.5 mL/min
   c. 30 mL/hr
   d. 2 mg/min (given)
   e. 120 mg/hr

251. a. 20 mcg/mL (0.02 mg/mL)
   b. 7.5 mL/min
   c. 450 mL/hr
   d. 150 mcg/min (given)
   e. 9000 mcg/hr (or 9 mg/hr)

252. a. 0.1 unit/mL
   b. 1 mL/min
   c. 60 mL/hr
   d. 0.1 units/min
   e. 6units/hr (given)

253. a. 100 mU/mL (0.01 unit/mL)
   b. 0.1 mL/min
   c. 6 mL/hr
   d 10 mU/min (given)
   e. 600 mU/hr

254. a. 5000 mcg/mL (5 mg/mL
   b. 2 mL/min
   c. 120 mL/hr
   d. 100 mcg /kg/min ×100 kg = 10,000 mcg/min
   e. 60,000 mcg/hr (60 mg/hr)

255. a. 10,000 mcg/mL (10 mg/mL)
   b. 1 mL/min
   c. 60 mL/hr
   d. 200 mg/kg/min × 50 kg = 10,000 mcg/min
   e. 600,000 mcg/hr (60 mg/hr)

256. a. 200 mcg/mL (0.20 mg/mL)
   b. 0.25 mL/min
   c. 15 mL/hr
   d. 50 mcg/min (given)
   e. 3000 mcg/hr (3 mg/hr)

257. a. 0.4 mg/mL
   b. 0.125 mL/min
   c. 7.5 mL/hr
   d. 0.05 mg/min
   e. 3 mg/hr (given)

258. a. 1000 mcg/mL (1 mg/mL)
   b. 0.4 mL/min
   c. 24 mL/hr
   d. 5 mcg/kg/min × 80 kg = 400 mcg/min
   e. 24,000 mcg/hr (24 mg/hr)

259. a. 500 mcg/mL (0.5 mg/mL)
   b. 2 mL/min
   c. 120 mL/hr
   d. 10 mcg/kg/min × 100 kg =1000 mcg/min
   e. 6,000 mcg/hr (6 mg/hr)

260. a. 0.6 mg/mL
   b. 0.166 mL/min
   c. 10 mL/hr
   d. 0.104 mg/min
   e. 6.25 mg/hr (given)

261. a. 50 units/mL
b. 0.4 mL/min
c. 24 mL/hr
d. 20 units/min
e. 1,200 units/hr (given)

262. a. 0.5 mg/mL
b. 0.166 mL/min
c. 10 mL/hr
d. 0.8 mg/min
e. 5 mg/hr (given)

263. a. 4 mg/mL
b. 0.06 mL/min
c. 4 mL/hr
d. 0.26 mg/min
e. 16 mg/hr (given)

264. a. 4 mg/mL
b. 0.5 mL/min
c. 30 mL/hr
d. 2 mg/min (given)
e. 120 mg/hr

265. a. 4 mg/mL
b. 0.75 mL/min
c. 45 mL/hr
b. 3 mg/min (given)
e. 180 mg/hr

266. a. 1600 mcg/mL (or 1.6 mg/mL)
b. 0.23 mL/min
c. 14 mL/hr
d. 5 mcg/kg/min × 75 kg = 375 mcg/min
e. 22,500 mcg/hr (or 22.5 mg/hr)

267. a. 100 units/mL
b. 16 mL/min
c. 10 mL/hr
d. 16.6 units/min
e. 1000 units/hr (given)

268. a. 200 mcg/mL (0.2 mg/mL)
b. 0.125 mL/min
c. 7.5 mL/hr
d. 25 mcg/min (given)
e. 1500 mcg/hr (or 1.5 mg/hr)

269. a. 0.4 units/mL
b. 0.5 mL/min
c. 30 mL/hr
d. 0.2 units/min (given)
e. 12 units/hr

270. a. 8 mg/mL
b. 0.375 mL/min
c. 22.5 mL/hr
d. 3 mg/min (given)
e. 180 mg/hr

271. a. 100 mcg/mL (0.1 mg/mL)
b. 2.1 mL/min
c. 126 mL/hr
d. 3 mcg/kg/min × 70 kg = 210 mcg/min
e. 12,600 mcg/hr (or 12.6 mg/hr)

272. a. 8 mg/mL
b. 1.87 mL/min
c. 112 mL/hr
d. 15 mg/min (given)
e. 900 mg/hr

273. a. 100 units/mL
b. 0.23 mL/min
c. 14 mL/hr
d. 23.3 units/min
e. 1400 units/hr (given)

274. a. 200 mcg/mL (0.2 mg/mL)
b. 2 mL/min
c. 120 mL/hr
d. 5 mcg/kg/min × 80 kg = 400 mcg/min
e. 24,000 mcg/hr (or 24 mg/hr)

275. a. 16 mcg/mL (0.016 mg/mL)
b. 0.125 mL/min
c. 7.5 mL/hr
d. 2 mcg/min (given)
e. 120 mcg/hr

276. a. 1 mg/mL
b. 0.083 mL/min
c. 5 mL/hr
d. 0.083 mg/min
e. 5 mg/hr (given)

277. a. 0.2 mg/mL
 b. 0.66 mL/min
 c. 40 mL/hr
 d. 0.133 mg/min
 e. 8 mg/hr (given)

278. a. 0.4 mg/mL
 b. 0.83 mL/min
 c. 50 mL/hr
 d. 0.33 mg/min
 e. 20 mg/hr (given)

279. a. 0.8 mEq/mL
 b. 0.2 mL/min
 c. 12.5 mL/hr
 d. 0.16 mEq/min
 e. 10 mEq/hr (given)

Ob/Gyn IV Drugs

280. a. $\frac{10 \text{ units/mL}}{500 \text{ mL}} = 20 \text{ mU/mL } (0.02 \text{ units/mL})$
 b. $20 \text{ mU/min} \times 60 \text{ min/hr} = 1200 \text{ mU/hr}$
 c. $\frac{1200 \text{ mU/hr}}{20 \text{ mU/mL}} = 60 \text{ mL/hr}$

281. a. 40 mU/mL (0.04 units/mL)
 b. 1200 mU/hr
 c. 30 mL/hr

282. a. 40 mU/mL
 b. 600 mU/hr
 c. 15 mL/hr

283. a. 40 mU/mL
 b. 900 mU/hr
 c. 22.5 ≈ 23 mL/hr

284. a. 10 mU/mL (0.10 units/mL)
 b. 60 mU/hr
 c. 6 mL/hr

285. a. 10 mU/mL
 b. 240 mU/hr
 c. 24 mL/hr

286. a. 10 mU/mL
 b. 720 mU/hr
 c. 72 mL/hr

287. a. 10 mU/mL
 b. 960 mU/hr
 c. 96 mL/hr

288. a. 10 mU/mL
 b. 120 mU/min
 c. 2 mL/min

289. a. 10 mU/mL
 b. 60 mU/min
 c. 1 mL/min